Consciousness, Biology and Fundamental Physics

Simon Raggett

authorHOUSE®

AuthorHouse™
1663 Liberty Drive
Bloomington, IN 47403
www.authorhouse.com
Phone: 1-800-839-8640

First published by AuthorHouse   01/26/2012

ISBN: 978-1-4678-8439-6 (sc)
ISBN: 978-1-4678-8440-2 (ebk)

Printed in the United States of America

Any people depicted in stock imagery provided by Thinkstock are models, and such images are being used for illustrative purposes only.
Certain stock imagery © Thinkstock.

This book is printed on acid-free paper.

CONTENTS

INTRODUCTION

In writing a book of this kind, it is difficult know what level to pitch it at and what degree of detail to bring in. On the one hand, experts in particular fields may ridicule the superficial nature of the description and arguments here, while at the other extreme some would-be readers may find even the opening sentences baffling. I have two recommendations for dealing with these problems. Firstly, I would advocate a pick and mix approach to the offerings here. For instance, those not particularly inclined to wade through user-unfriendly material relative to physics, biology and neuroscience might prefer to go straight to the final section 6, rather arrogantly entitled 'A Theory of Consciousness'. This gives the main conclusions as to how consciousness might arise and its function. If this looks at all interesting it is then possible to go back and see how I have attempted to substantiate the proposals have made in this section.

The same general approach can be applied to the other chapters, in skipping over things that are either too difficult, or are too well known to need revisiting. The first few pages could be comfortably missed as they give an outline sketch of conventional consciousness ideas, and the objections to them. I do not believe that these theories have any explanatory value and it might be possible to skip to the 'Consciousness as a Fundamental Theory' a few pages later.

There is perhaps a word of caution relative to this approach. The section on physics emphasises the problem areas in quantum physics, which may be played down in more mainstream discussion. The sections on both quantum biology and neuroscience emphasise research work in very recent years that can be argued to have reversed some assumptions that are still commonly held in consciousness studies.

The detailed structure of the book may also require some explanation or even excuse. As I am neither a physicist, neuroscientist or biologists, and as this area is often the target of dismissive, if badly researched, accusations of pseudoscience, I have relied in many places on summarising the work of affiliated and usually peer-reviewed researchers. This can involve a degree of overlap or repetition, which may be tedious, but should be taken as demonstrating the scientific pedigree of various ideas. This is particularly applicable to the quantum biology section.

The main inspiration for this attempt at consciousness theory are the ideas of Roger Penrose (1.& 2.). Unfortunately, I have over more than twenty years come to the opinion that the vast majority of modern consciousness studies is profoundly misguided, and that in time Penrose may come to be seen, as being alone as a deep thinker on the subject, in our benighted period.

The dark night of the mind: Another thing for which I will not apologise is that much of this book may seem difficult. One thing that does often amaze me in consciousness studies is the proportion of people who expect to come on a quick easy solution. Inspite of a huge outpouring of books and papers from the scientific and philosophical communities in the last two decades, we appear no closer to a convincing consensus theory. David Chalmers called consciousness the hard problem, and this looks even more justified than when he coined this description back in the 1990s. If we do want to establish a theory of consciousness, it looks very likely that we will have to come to a hard place in terms of unravelling biology, neuroscience and probably physics.

This book attempts an amendment and simplification of the Penrose/ Hameroff Orch OR scheme, and also an attempted updating in line with very recent developments in biology. It is tentatively suggested that a less complex approach to the function of consciousness than that provided by the Gödel theorem can be attempted, and similarly that in the brain, quantum consciousness might be based on shorter-lived quantum coherence in individual neurons, rather than the longer-lived and spatially distributed proposal put forward by Hameroff. The possible need to amend the original concepts are the reason for my moving from merely commenting on quantum consciousness topics to outlining a more personal version of the consciousness theory.

SECTION 1

THE HARD PROBLEM OF CONSCIOUSNESS

DEFINITION OF CONSCIOUSNESS: Consciousness is defined here as our subjective experience of the external world, our physical bodies, our thinking and our emotions. Consciousness is also defined in terms of it 'being like something' to exist and have experiences. It is like something to be alive, like something to have a body and like something to experience the colour red. In contrast, it is assumed to be by most not like something to be a table or a chair. Further to being like something conscious also gives us the experience of choice or preferences. In philosophy, this opens up the controversial topic of freewill, but at a more mundane level we have the something it is like to choose types of beer, or between a small amount of benefit now or a more substantial benefit in the future. Another special characteristic of subjective consciousness is privacy, in the sense that we have no way of knowing that our experience of the colour red is the same as someone else's, and no way of conveying the exact nature of our experience of redness. These subjective experiences are referred to as qualia. The problem of qualia or phenomenal consciousness is here viewed as the whole of the problem of consciousness.

THE HARD PROBLEM

The problem we have to address here is how consciousness, subjective experience or qualia arise in the physical matter of the brain. Even this simple question raises some queries as to whether consciousness does in fact arise from the brain, although the arguments in favour of this position do in fact look strong. The classic argument is that things done to the brain such as particular injuries or the application of anaesthetics can remove consciousness.

DUALISM

The main challenge to the 'brain produces consciousness' hypothesis is dualism, or the idea that there is a separation between a spirit stuff and a physical stuff that together make up the universe, with consciousness being part of the spirit stuff, but inhabiting a physical brain and body. This had probably been the most popular idea since ancient times, but it was formalised by Descartes in the seventeenth century. The idea has a certain beguiling simplicity, since at a stroke it gets rid of the need to worry about how the physical brain can produce consciousness, or all the difficulties this gives rise to in terms of biology and physics.

Unfortunately, the problems of dualism appear to be of the serious kind. This is principally the question of how the physical stuff and the spirit stuff can interact. If the spirit stuff is to interact with the physical stuff it would appear to need to have some physical qualities, in which case it would not be true spirit stuff. The same applies in the opposite direction in that the physical stuff would seem to need some spirit qualities to interact with the spirit stuff, and would therefore not conform to conventional physics.

We are thus left with the problem of how the physical stuff as described by science can produce consciousness. The philosopher, David Chalmers, labelled this the 'hard problem' (3.). The problem here is really a problem of specialness. The brains of humans and possibly animals are the only places in the universe where consciousness has been observed, so the question really is as to 'what is special about the brain', and the answer to this tends to be that there's nothing special about the brain, because it's made of exactly the same type of stuff and obeys the same physical laws as the rest of the universe. The brain comprises the same carbon, oxygen, hydrogen and other atoms that are found in the stars and planets and the objects of the everyday world around us.

At first sight this might not seem too much of a problem. The brain is considered to be the most complex thing in the universe, and surely something in such a system can manage to produce consciousness. Unfortunately, this does not appear to be the case. In a conventional neuroscience text book, which will emphasise the fluctuation of electrical potential in the neurons (brain cells) and the resulting movement of neurotransmitters between neurons, we are presented with a causally closed information system, which does not require consciousness in order to function, and offers no physical mechanism by which consciousness could be produced.

Since consciousness ceased to be a taboo subject for academic research twenty-or-so years ago, several theories that seek to explain how consciousness could arise within classical/macroscopic physics have been advanced. It would take many hundreds of pages to discuss these adequately, so I will here summarise the main ideas, and where they look to fail. For those who find any of them plausible, there is a huge and expanding literature out there working to reinforce these theories.

CONSCIOUSNESS AS AN EMERGENT PROPERTY

Possibly the most plausible attempt to explain how the brain could produce consciousness within the concepts of classical and macroscopic physics is the idea that consciousness is an emergent property of the brain's physical matter. Emergent properties are an established concept in physics. The classic example is that liquidity is an emergent property of water. The individual hydrogen and oxygen atoms or their sub-atomic components do not have the property of liquidity. However when the atoms are combined into molecules and a sufficient number of these are brought together, within a particular temperature range, the property of liquidity emerges.

The problem with the emergent property idea, when it comes to dealing with consciousness, is that where emergent properties do arise in nature, physics can trace them back to the component particles and the forces that bind them. Thus the liquidity of water can be explained by the electromagnetic force acting on hydrogen and oxygen atoms. But in many years of the emergent property idea being promoted by parts of the conventional consciousness studies community, no one has been able to propose a microscale emergent mechanism for consciousness in the brain, comparable to the explanation of how liquidity emerges from water.

FUNCTIONALISM

In much of the late twentieth century consciousness studies was dominated by functionalism. This theory proposes that consciousness is a function of the brain's information processing system, and that the biological matter of the brain is irrelevant to consciousness. This means that any system that processes information in the same way as the brain will be conscious regardless of what it is made of. Therefore a silicon computer of sufficient complexity would flip into consciousness at some point, and future systems using still other materials would do likewise. This is because the system, rather than the stuff from which it is made, is seen as being the thing that produces consciousness. The underlying weakness of functionalism is that it does not actually explain the mechanism of how consciousness arises in the brain's systems in the first place, nor how it might physically arise from silicon computers or other machines.

This is a crucial problem regardless of whether the brain or system in question is made of biological tissue, silicon or anything else. It is generally agreed that the computer on the desk is not conscious, but that brains are conscious. The question we are left with is what changes between the computer on the desk and the brain, and similarly between the computer on the desk and any future super computer that might actually become conscious. There may be a vague assumption that more and more of the same initial complexity does it. But the physical world doesn't work like that. The problem of butter not cutting steel is not resolved by adding lots more butter, but by finding something with different electromagnetic properties from butter.

IDENTITY THEORY

Identity theory is similar in tone to functionalism. It says that consciousness is identical to the brain or at least parts of it. The problem with an identity theory is that it needs to specify a particular object, or more plausibly a particular process in the brain that is physically identical to consciousness. It is not enough to show that the axons of neurons spike, or that there is a gamma oscillation between the cortex and the thalamus, when conscious processing occurs. These things are correlated to consciousness, but that is another thing from saying they are identical to consciousness. The distinction between identity and correlation is crucial here. Thunder and lightning are correlated, but they are not the same physical process, even though they have the same ultimate cause. In contrast, the morning star and the evening star are identical, because they are both names given to the planet Venus, a single physical object. Astronomy has conclusively demonstrated this identity, because the behaviour of a point of light in the morning and evening sky can be completely explained by what we know about this planet Venus. However, neuroscience has not demonstrated that the behaviour of any particular physical process in the brain that is identical to, or can completely explain the behaviour of consciousness, as opposed to being merely correlated to it.

In addition to this, more recent neuroscience has at least qualified identity theory. Expositions of identity theory tended to be rather simplistic in applying to the whole brain, while recent neuroscience has demonstrated consciousness as correlated to both particular neuronal assemblies and to single neurons, albeit on a temporary basis with activity correlated to consciousness shifting from place to place in the brain.

HIGHER ORDER (HOT) THEORIES

The basic idea here appears to be that a level or perhaps levels of the brain observe another level or levels, and the interaction of the two somehow generates consciousness. We are essentially also asked to believe that because one system monitors another it will become conscious. This hypothesis bears little relation to the technological world where it is common place for one non-conscious automated system to monitor another and have some automatic response to changes in it, without any requirement for consciousness.

EMBODIMENT AND CONSCIOUSNESS

In the present century, the concept of conscious embodiment has come to the fore. It is suggested that a brain or a computer by itself cannot be consciousness, but the brain and possibly the computer can become conscious when attached to a body or some comparable extension.

The recognition of the fact that brain and body are interactive was in itself an advance on twentieth century notions of the brain as an isolated computer, and the body as an automaton incapable of being uninfluenced by the mind. That said, there appear to be two problems with this approach as an explanation of consciousness. Firstly, it carries the rather implausible notion that the body has some consciousness generating process that does not exist in the brain. There is a complete absence of explanation as to what this might actually be. Admittedly, most touch and pain are transmitted from the body to the brain, and visual and auditory inputs to the brain are fed forward to the viscera, but this does not explain why signals going through the body should generate something different from incoming signals through the brain.

Further, this theory looks difficult to square with what has now become known about the organisation of brain processing. While the bodily touch and pain can certainly be seen to play a role, it is hard to see why all visual, auditory inputs, and the results of cognitive processing should have to wait on the laborious responses of the viscera, especially as it is the reward assessment areas of the brain that signal the viscera in the first place. If bodily generated emotion were the whole story, the emotional evaluation regions of the orbitofrontal and amygdala would seem to be in a state of suspended activity between sending a signal to the autonomic system and getting signals back from the viscera. In the specific case of rapid phobic reactions in the amygdala, the idea fails completely. Recent expositions of the theory indicate an over emphasis on the body's movement and relations to the external world, perhaps because they are more compatible with this theory, at the expense of the other senses, and more especially at the expense of thinking and emotion—related evaluation.

A further objection to this theory is that bodily arousal does not provide a sufficient range of responses to match the range of human emotional responses. Emotional research, which often means animal research, has tended to focus on the easy target of fear, which produces very definite bodily responses, whereas cognitive processing or visual and auditory sensations, not related to immediate danger, can produce a much less marked bodily response, and a wider and subtler range of emotional responses. The more plausible view is that visceral responses are one aspect of many responses that are integrated in the orbitofrontal and other evaluation processes.

Further to this, evolution seems to have altered the response system to visceral inputs when it came to primates. The visceral inputs no longer go via the pons structure in the brain stem, and this is argued to suggest a less automatic response to visceral inputs in humans and primates. It seems more likely that in line with most brain processes, there is a complex feed-forward and feedback between all parts of the system including the viscera and the orbitofrontal. The body-only theory looks to depend on a simple feed-forward mechanism, which is alien to how brain processing functions.

INFORMATION THEORY AND CONSCIOUSNESS

Attempts to classify consciousness as a form of information can be seen as another attempt to explain consciousness in classical terms. This idea also looks to encounter insuperable problems. There are innumerable examples of information processing and communication that does not involve consciousness, especially when we look at modern technologies. Further to this, we lack a description of a physical process that would distinguish conscious information from non-conscious information.

There is a core difference between information and reality, in that information involves only what we happen to know about something, while a knowledge of reality requires a full description of its make up and a full explanation of its behaviour. The only information available to a hunter gatherer in ancient Africa glancing up at the sun is the intensity of glare and heat and the changes in its position in the sky. It required the complexities of modern science to unravel everything that is involved in the sun producing light, the light getting to our eyes, and the brain states this produces.

EPIPHENOMENALISM

This theory proposes that consciousness is a by-product of neural processing, which has no function or significance. There are three main problems here. In the first place, as with some other modern consciousness theories, it is actually a non-explanation. Even if consciousness has no function, we still need to know how it is physically instantiated, and this is never attempted when this theory is proposed.

The suspicion is that the proponents of this theory are unconsciously closet Cartesians, with an underlying assumption that consciousness is 'non-physical' or 'immaterial.' If it can be categorised in this surprisingly unscientific manner it can be dismissed as non-functional, and relegated to a smallest possible footnote in any scientific study. This is contradictory in that the proponents are invariably non-dualists, who believe that there is no such thing as the non-physical.

The second problem is that consciousness has to be linked to the rest of the brain and the physical universe, because the very fact of conscious experience indicates that we are dealing with the reception of some form of incoming signal, and anything receiving incoming signals is likely to be able to emit them in some form of response, which will have physical consequences. Some writers have suggested an escape route here, which allows consciousness a trivial influence. This is feasible up to a point, but hints at problems in defining what is trivial, and would erode the position of the modern orthodoxy that argues for complete determinism and no freewill at all in behaviour.

A further problem for epiphenomenalism is that it conflicts with evolutionary theory. If this by-product consciousness is physical as the scientific paradigm demands, it needs energy to produce and maintain it, and given that the brain is very energy intensive, this could involve quite a large amount of energy. It would be maladaptive for evolution to select for something that ties up energy with no benefit to the organism. It might be argued that neural processing was such that some by-product was essential, but this would require a demonstration that neural processing produces this something else. However, in the physical description of the matter and energy involved in the brain, as described by standard neuroscience, there is no sign of such a process.

NEW MYSTERIANS

New mysterians or sometimes just mysterians take the view that just as dogs cannot understand calculus, humans will never be able to understand consciousness. This may in the end turn out to be true, but to accept this view as final at this stage in the proceedings seems unduly defeatist. The human mind has proved capable of understanding the mechanisms of the physical universe so far, and it is reasonable to hope that the rather narrow scope of thinking in conventional consciousness studies may not have exhausted all possible explanatory routes.

Where the mysterian approach is advocated there is usually a 'no nonsense' implication that having established this point, consciousness is no longer a threat to a view of the mind that is dominated by classical physics and slightly old fashioned text book neuroscience. On further reflection however the exact opposite is true. Humans have been able to understand the physical law. If they cannot understand consciousness, then consciousness lies outside the physical law or any logical extension to it. This, if anything ever does, opens the sluice gates to the dark tide of the occult, necessitating that consciousness is something akin to a spirit stuff lying outside of, and able to act outside of the physical law.

IS THE PROBLEM WITH CONSCIOUSNESS STUDIES CULTURAL RATHER THAN SCIENTIFIC?

Much of the scientific, philosophical and psychological community never internalised the revolution in physics that produced quantum theory early in the last century. There seems to be an assurance that this is an abstruse special case that need not bother day-to-day thinking. The theory was more or less censored out of general education and even basic scientific education. As a result, a purely Newtonian view of the world has become the 'common sense' norm for people who know little about either classical or quantum physics. In mainstream consciousness studies, there is an apparent determination not to move beyond nineteenth century macroscopic physics, which proposes a billiard ball world, where everything is explained in terms of objects bumping into one another. This is despite the fact that it has been known for a century that this is a convenient approximation for studying the human-scale world, but is not how the underlying physics works.

Neuroscience's approach to consciousness is even more mired in nineteenth century concepts. The discovery of individual neurons and their connections at the end of that century allowed the idea of the neuron as a simple switch with no further complications to become entrenched. Not long after this discovery, what is sometimes called 'the long night of behaviourism' descended on consciousness studies, decreeing that consciousness was irrelevant to behaviour and not a proper subject of study. Although behaviourism as such dropped out of favour in the latter decades of the twentieth century, subsequent theories have sought to justify the same general conclusion by marginalising consciousness. Behaviourism is dead. Long live behaviourism.

In fact, one curious consequence of the functionalist and identity approaches is that much of consciousness studies has paid remarkably little attention to the brain or to advances in neuroscience in recent decades. The assumption has been that all that was needed was a particular system that could run on any material, and there was no need to inquire any further into the detailed biology of the brain. Information about binding and the gamma synchrony or consciousness in individual neurons and the distinction between conscious and non-conscious neural area are footnotes, while the functional role of subjectivity in orbitofrontal valuation is never mentioned, or perhaps not even known about.

WHY 21ST CENTURY CONSCIOUSNESS STUDIES WILL FAIL

Consciousness studies has gone off in a different direction from neuroscience. Much of it is dominated by philosophers or psychologists who deal more in abstractions than what is going in the physical brain. In addition, they have tended to see themselves as under-labourers supporting a nineteenth century Newtonian world view, while at the same time discussing consciousness in very abstract terms that take limited account of advances in neuroscience research.

Neuroscientists, meanwhile, seem to have been pressured into treating consciousness as not really part of their remit, and defer to philosophers whenever they felt it necessary to mention consciousness, even when the views of the philosophers appeared to conflict with the neuroscientists own more recent findings. For this reason, it seems possible to predict that consciousness studies will come to the end of the 21st century without having achieved consensus on a theory that has any useful explanatory value.

CONSCIOUSNESS AS A FUNDAMENTAL PROPERTY

The above discussions might seem to bring us to an impasse, where we don't think that consciousness can derive either from separate spirit stuff nor from the material that comprises the brain, the body and the universe. Luckily, there is an escape route from this. Physics does not explain everything. The arrow of explanation heads for ever downwards, but it does at last strike bottom. There is a level beyond which there is no further reduction or explanation. The quanta or sub-atomic particles have properties of mass, charge and spin and are bound by the particular strengths of the forces of nature. These are fundamentals, primitives or given properties of the universe that have to simply be accepted.

If we ask what is the charge on the electron, not what does it do, but what is it, the answer is a resounding silence, because it is a given property of the universe, and comes without explanation. If we had a scientific culture that did not accept that quanta could be electrically charged, and that other quanta could intermediate the electromagnetic force, this might develop into another hard problem like the one we have with consciousness. We would go round and round trying to stick electrical charge on to other and probably macroscopic physical features, or we might even decide, as happens sometimes in consciousness studies, that charge did not really exist, that it was a product of something else or an illusion. No doubt experimental psychologists could devise cunning tests that showed how subjects confabulated the idea of electrical charge.

If we accept that fundamental properties do exist, and that they cannot be explained by other means, and also that it is impossible to explain consciousness in terms of classical physics, then it would seem reasonable to suggest that consciousness is one of this small group of fundamentals. Thinkers such as David Bohm (4.) and Roger Penrose have made such proposals, but the response has been generally hostile, although the reasons for this may be cultural or even metaphysical rather than scientific.

SECTION 2

QUANTUM PROBLEMS & THE NATURE OF SPACETIME

Just having a concept of consciousness as a fundamental of physics is not by any means enough. Fundamental physics may be a possible gate to consciousness, but to substantiate this we need some concept of how consciousness might be integrated into what is known about fundamental physics. In the first place, it might help to have at least a very simplified idea of quantum theory and some recent ideas about spacetime.

Quantum theory is the fundamental theory of energy and matter as it exists behind the appearances of the classical or macroscopic world. Suppose one were to ask for a scientific description of your hand. Biology could describe it in terms of skin, bone, muscles, nerves, blood etc., and this might seem a completely satisfactory description. However, if you were just a bit more curious, you might ask what the muscle and blood etc. were made of. Here you would descend to a chemical explanation in terms of molecules of protein, water etc. and the reactions and relations between these.

If you were still not satisfied, then beyond this you would have to descend into the quantum world. At this level, the solidity and continuity of matter dissolves. The molecules of protein etc. are made up of atoms, but the atoms themselves are mainly vacuum. Most of the mass of the atom lies in a small nucleus, comprised of protons and neutrons, which are themselves made up of smaller particles known as quarks. The rest of the mass of the atom resides in a cloud of electrons around the nucleus.

THE FORCES OF NATURE

The fundamental particles are bound together by the four forces of nature, which are electromagnetism, the strong and the weak nuclear forces and gravity. The quanta can be divided into two main classes, the fermions, which possess mass and the bosons which convey energy or the forces of nature. In contrast to the nuclear forces, gravity and electromagnetism are conceived of as extending over infinite distance, but with their strength diminishing according to the inverse square law. That is, if you double your distance from an object, its gravitational attraction will be four times as weak. The strong nuclear force binds together the particles in the nucleus of the atom, and acts only over the very short range. Gravity is a long-range force that mediates the mutual attraction of all objects possessing mass. The electromagnetic force, also a long-range force, is perhaps the force most apparent in everyday life. We are familiar with it in the form of light, radio, microwaves and X-rays. It holds together the atom through the attraction of the opposite electrical charges of the electron and the proton. It also governs the interactions between molecules. Van der Waals forces, a weak form of the electromagnetic force is vital to the conformation of protein and thus to the process of life itself.

Quantum waves, superpositions and a problem of the serious kind

The quantum particles or quanta are unlike any particles or objects that are encountered in the large-scale world. When isolated from their environment, they are conceived as having the property of waves, but when they are brought into contact with the environment, there is a process referred to as decoherence, wave function collapse or measurement, in which the wave form becomes a particle located in a specific position.

The wave form of the quanta is different from waves of matter in the large—scale world, such as the familiar waves in the sea. These involve energy passing through matter. By contrast, the quantum wave can be viewed as a wave of the probability for finding a particle in a specific position. This probability wave also applies to other states of the quanta such as momentum. While the quanta remain in wave form, they are described as being in a superposition of all the possible positions that the particle could occupy. At the peak of the wave, where the amplitude is greatest, there is the highest probability of finding a particle. However, the choice of position for each individual particle is completely random, representing an effect without a cause. This acausal result comprises the first serious conceptual problem in quantum theory.

THE TWO-SLIT EXPERIMENT

The physicist, Richard Feynman, said that the two-slit experiment contained all the problems of quantum theory. In the early nineteenth century, an experiment by Thomas Young showed that when a light source shone through two slits in a screen, and then onto a further screen, then a pattern of light and dark bands appeared on the further screen, indicating that the light was in some places intensified, and in others reduced or eliminated. Where two waves of ordinary matter, for instances waves in water, come into contact an interference pattern forms, by which the waves are either doubled in size or cancelled out. The appearance of this phenomenon in Young's experiment demonstrated that light had the characteristics of a wave.

THE EXPERIMENT REFINED

Later, the experiment was refined. It could now be performed with one or two slits open. If there was only one slit open, the photons or light quanta, or any other quanta used in the experiment behaved like particles. They passed through the one open slit, interacted with the screen beyond, and left an accumulation of marks on that screen, signifying a shower of particles rather than a wave. But once the second slit was opened, the traditional interference pattern, indicating interaction between two waves, reappeared on the screen. The ability to generate the behaviour of either particles or waves, simply according to how the experiment was set up, showed that the quanta had a perplexing wave/particle duality.

It could seem that the best way to understand what was happening here was to place photon counters at the two slits in order to monitor what the photons were up to. However, as soon as a photon is registered by a counter, it collapses from being a wave into being a particle, and the wave-related interference pattern is lost from the further screen. The most plausible way to look at it may be to say that the wave of the photon passes through both slits, or possibly that it tries out both routes.

THERE WAS WORSE TO COME

The wave/particle duality was shocking enough, but there was worse to come. Technology advanced to the point where photons could be emitted one-at-a-time, and therefore impacted the screen one-at-a time. What is remarkable is that with two slits open, but the photons emitted one-at-a-time, the pattern on the screen formed itself into the light and dark bands of an interference pattern. The question arose as to how the photons emitted later in time 'knew' how to arrange themselves relative to the earlier photons in such a way that there was a pattern of light and dark bands. The ability of quanta to arrange themselves in this non-random way over time, despite initially choosing random positions, could be considered to be the second big problem of quantum theory.

QUANTUM ENTANGLEMENT

Einstein disliked the inherent randomness involved in the collapse of the wave function. This was despite himself having contributed to the foundation of the quantum theory. He sought repeatedly to show that quantum theory was flawed, and in 1935 he seemed to have delivered a masterstroke in the form of the EPR (Einstein, Podolsky, Rosen) experiment. At the time, this was only a 'thought experiment', a mental simulation of how a real experiment might proceed, but in recent decades, it has been possible to perform this as a real experiment.

Two quanta that have been closely connected can be in a state where they will always have a particular relationship to one another. This is known as being entangled. For instance, electrons have a property of spin, and can have a state of spin-up or spin-down. For example, two entangled electrons can be in a state where their spin will always be opposite. This applies however distant they become from one another. However while the electrons (or other quanta) are in the form of the wave, both electrons are superpositions of spin-up and spin-down, so entanglement only really manifests itself when there is decoherence or wave function collapse.

The EPR experiment proposed that two quanta, which have remained sufficiently isolated from their environments to be conceived as waves or superpositions, are moved apart from one another. This could be a few metres along a laboratory bench or to the other side of the universe. The relevant consideration is that the two locations should be out-of-range of a signal travelling at the speed of light, within the timescales of any observations that are being made.

Both particles are a superposition of two possible states, but if an observation is made on one of the particles, its wave function collapses, and it acquires a defined spin, let's say spin-up in this case. Now when an observation is made on the other particle, it will always be found to have the opposite spin. This defies the normal expectation of classical physics that a random choice of spins would produce approximately 50% the same spin and 50% different. Therefore, there is seen to be some non-local connection between the two particles, although it is not possible to describe or detect this in terms of a physical transfer of energy or matter. In fact, the entanglement influence is shown to be instantaneous, whereas energy and matter are thought to be constrained by the speed of light. This quantum relationship between particles is called entanglement, and can be regarded as the third big problem in quantum theory.

QUANTUM INTERPRETATIONS

Recent debate suggests that rather than showing any sign of moving towards any kind of consensus, the different interpretations of quantum theory are becoming more distinct and more entrenched (5.). In particular, six types of approach are distinguished, [1.] Everett many-world theories [2.] Post-Copenhagen theories based only on our information about quantum states. [3.] Coherence remains with hidden superposition within macroscopic objects. [4.] Bohmian type pilot-wave theories, [5.] wave function collapse theories. [6.] The suggestion that none of these are satisfactory, and that quantum theory will only be explained in terms of a deeper level of physics.

The interpretation of quantum theory has an unhappy history. In the 1920s there was for a short time a unity of purpose in trying to both understand and apply quantum theory. Thereafter a premature notion that the interpretative debates had been settled took hold, and in the period after World War II academic institutions discouraged foundational research. The physicist, AntonyValentini, argues that quantum theory got off to this bad start, because it was philosophically influenced by Kant's idea that physics reflected the structure of human thought rather than the structure of the world. The introduction of the observer into physics allowed a drift away from the idea of finding out about what existed and also how what existed behaved. It was not until the 1970s and 1980s that new interpretations of quantum theory started to become academically acceptable.

The philosopher, Tim Maudlin, contrasts two intellectual attitudes in the approach to quantum theory. Einstein, Schrödinger and Bell wanted to understand what existed and how it worked, while many who came after them were more incurious, and happy with a calculational system that worked. This is the so-called 'shut up and calculate' approach. Maudlin suggests that what is traditionally referred to as the 'measurement problem' in quantum theory is really the problem of what is reality. He sees the aim of physics as being to tell us what exists and the laws governing the behaviour of what exists.

Maudlin argues that quantum theory describes the movement of existing objects in spacetime, while the wave function plays a role in determining how objects behave locally. He suggests that there are many problems for theories that deny the existence of real objects or the reality of the wave function. Lee Smolin, a physicist at the Perimeter Institute, remarks that bundles of ideas in quantum theory and related areas tend to go together. Believers in Everett many worlds tended to also support strong artificial intelligence, allowing classical computers to become conscious, and also support the anthropic principle. Disagreement with these three ideas also seems to go together.

EVERETT MANY-WORLDS

The philosopher, Arthur Fine, puts the fashionable 'many worlds' theory, originally proposed by Everett, at the bottom of 'anyone's list of what is sensible'. He criticises proponents of the theory for concentrating on narrow technical issues rather than thinking about what it means for universes to split. The difficulties of many worlds may be even greater than Fine's suggests. The splitting of worlds demands that huge number of new universes are coming into existence all the time, thus apparently suggests that the energy of entire new universes is being created the whole time. Explanations never seem to go beyond asserting that this is, for some reason, not a problem. Christopher Fuchs, another Perimeter Institute physicist, criticises philosophers who support the Everett theory for not looking for some physical explanation. The theory did not receive much support when it was originally propounded in the 1960s. The current popularity may look like an attempt to preserve classical assumptions, even at the cost of asserting a fantastical sci-fi idea.

INFORMATION THEORY

Shelley Goldstein, who spans maths, physics and philosophy, criticises information based theories for their failure to deal with the two-slit experiment. He asks how the different paths of the wave function in the two-slit experiment could lead to a wave interference pattern if nothing physical was involved. He thinks that the wave function is objective and physical, and neither some form of purely subjective experience, nor something that is simply the information that we happen to have. He sees the notion of information more in terms of a brain state connected to human needs and desires, rather than as an objective aspect of the external world.

Goldstein discusses a refined version of the double-slit experiment, in which the quanta are sent into the system one-by-one and an interference pattern gradually emerges. He sees the emerging pattern as an entirely objective phenomena not resulting from a limitation on our knowledge of the system. Tim Maudlin appears to agree with this, arguing that in the two-slit experiment, the sensitivity to whether one or two slits are open indicates the response of something physical, rather than just the experimenters ignorance about the location of a particle. Maudlin points to the holistic nature of the two-slit experiment, and suggests that the same thing is apparent in non-locality.

The philosopher, David Wallace, also takes the view that states in physics are facts about the physical world, and not just our knowledge of the physical world. He rejects views that see the quantum as a mixture of our information and our ignorance, because in practise physicists measure particular physical processes. The physicist, Antony Valentini, poses the question as to how the definite states of classical physics arise from the indefinite states of quantum physics. He argues that it is impossible to have a continuous transition or emergent process moving from one to the other. The problem of measurement or reality at the quantum level is therefore argued to be a real problem, and requires some physical theory such as pilot waves or collapse theories to explain it.

The physicist, Ghirardi, a member of the trio of physicists responsible for the GRW collapse theory, views information theory as having played a negative role in terms of evading the need to deal with foundational problems in quantum theory. He sees it as a backward step to go from being concerned about what exists, to merely considering our limited information. John Bell, whose inequalities theorem sparked off the modern interest in entanglement, asked in response to this approach what the information was about. Proponents of information theory denounce this as a metaphysical question, which seems illogical for physicists who are themselves apparently withdrawing from the attempt to produce a physical description of the universe. As in some reaches of consciousness studies, we seem to be seeing the modern mind retreating into a mysterian view, possibly as a last ditch way of defending classical physics, or perhaps we should say metaphysics. Tim Maudlin similarly finds the notion of information theory puzzling. In his view the physical reality exists before we start to get information about it, and it is not meaningful to reverse this process.

DECOHERENCE THEORY

Tim Maudlin uses reductio ad absurdum to argue against decoherence theory. Buckyballs (a molecule of 60 carbon atoms) and some biomolecules have been put into superposition, and the line of decoherence argument suggests that larger and larger superpositions are possible without limit, so that decoherence never occurs and superpositions remain hidden in macroscopic objects.

From this Maudlin argues the solid macroscopic objects such as bowling balls should be capable of being put through a two-slit experiment and produce an interference pattern. I suppose the counter argument might be that the superposition is too small to produce that type of separation and interference, at least that's what decoherence arguments appear to be talking about. However, quantum superposition, as in a normal two-slit experiment, looks to imply complete separation, and this should be achievable by bowling balls or even iron cannon balls if decoherence theory is to be validated.

Similarly Ghirardi says that he would be willing to give up his collapse interpretation of quantum theory, if macroscopic superpositions could be demonstrated. Other philosophers also object to residual approximation and lack of explanation for superposition in the decoherence approach. Ghiradi is also critical of the theoretical basis of decoherence theory, where macroscopic objects are deemed to remain in superpositions, although these are superpositions that cannot be detected by existing technology. He sees this as reversing the normal process of science, and attempting to move from our definite knowledge of macroscopic objects to an approximation.

THE PILOT-WAVE MODEL

The 'pilot-wave' model developed by de Broglie in the 1920s, and was revived by Bohm in the 1950s. This can be argued to be the simplest solution to the measurement or reality problem. Particles existing in reality are argued to be guided by the wave function equation. There are perceived problems with this approach. The wave function itself is supposed to be real as well as the particle, although there is no evidence for this, and the evolution of the wave function without collapse is argued to have implications for a many worlds outcome. The limitation on the Bohmian approach is that it is impossible to know the initial conditions that would give control over the otherwise random outcome of a quantum experiment.

COLLAPSE MODELS

Wave function collapse models developed by Ghirardi and others are yet another interpretation of quantum theory. These theories require a modification of the Schrödinger equation, so that the evolution of the wave function described by the Schrödinger equation can collapse to the outcome of a particle with a particular position and other properties.

In Ghirardi's theory of wave function collapse, the wave function can be viewed as the quantity that determines the nature of our physical world and the spatial arrangement of objects. The wave function governs the space and time of the localisation of elementary particles. He prefers collapse theories that assume a process for random localisation of particles operating alongside the standard Schrödinger quantum evolution. Such localisation occur only rarely for quanta, but in a rigid body a localisation in one place will lead to localisation of the whole body. This is seen as defining the distinction between quantum and classical processes. Tim Maudlin defends the Ghiradi-Rimini-Weber (GRW) interpretation of quantum theory against the criticism that it violates conservation of energy. His approach is to accept that the theory does imply a small amount of heating up in the universe over time, but argues that this is not out of line with observation and preserves the general principle of conservation.

MATHEMATICAL SUPPORT FOR COLLAPSE THEORIES

As of November 2011 collapse theories look to have received a degree of support from researchers, Pusey, M., Barrett, J. & Rudolph, T. at Imperial College London (6.), who have devised a theorem claiming to prove the physical reality of wave function collapse. (arXiv: 1111.3328v1 [quant-ph] 14 Nov 2011 and nature.com) The authors claim to have shown by their theorem that the view that quantum states are only mathematical abstractions (referred to as the statistical interpretation) is inconsistent with the predictions of quantum theory, and that therefore quantum states are real physical states.

The theorem indicates that quantum states in an experiment must 'know' what state they have been prepared in, i.e. they must be physical systems, or an experiment will have results not predicted by quantum mechanics. They also claim that it is feasible for the theorem to be tested by experiment. Against this it should be noted that some commentators on nature.com argue that there are errors in the authors' work.

Schrödinger originally conceived the wave function as a physical state, but others soon argued that the wave function was not physical, or was merely a convenient fiction, a calculational procedure, or an encoding of experimenters limited information.

The view that the wave function was only a mathematical abstraction was the basis of the Copenhagen interpretation of quantum theory that dominated thinking through much of the 20th century. Support for the Copenhagen interpretation was eroded in the latter part of the 20th century, but the idea of the wave function as a mathematical abstraction has more recently been given a new lease of life by quantum information, which views the wave function as abstract information.

If the authors' theorem was to be vindicated it would not merely discredit quantum information theory but conclude the debate of nearly a century as to whether the quantum wave function was a physical reality. Clearly, this has crucial implications for Penrose and similar theories that look to the reality of the wave function to open physical access to a fundamental level of the universe.

DEEPER LEVELS

Lee Smolin (7. & 8.) himself takes a 'tragic' view of modern science, in thinking that it does not even have a toehold on why the universe appears quantum or on the nature of consciousness. Smolin meanwhile looks for a more fundamental theory. He thinks that space and the concept of locality should be thought of as emerging from a network of relationships between the quanta that are regarded as fundamental. He argues that the existence of non-locality shows that spacetime is not fundamental, in that non-locality does not accord with the conventional view of spacetime. He proposes that spacetime emerges from a more connected structure, with the non-local connections observed in quantum theory being a remnant of this more connected system. Lucien Hardy, another physicist at the Perimeter Institute, argues that while quantum theory has probabilities evolving against a fixed spacetime background, relativity is deterministic but has a dynamic version of spacetime. He thinks that some deeper theory of quantum gravity will need to combine the probabilistic and dynamic aspects of these two theories.

There is no easy conclusion to this debate, although the combination of the two-slit and EPR experiments seems to point to the quanta being bound into a non-local structure. A possible solution is that randomness, and the arrangement of the particles in the single-emission two slit experiment are alternative faces of entanglement, which describes the communication of quantum properties across spacetime. The approaches of many modern physicists tend to view spacetime as a discrete network or web, and this in turn hints at a structure which could support some form of pattern or code that could be related to subjective experience and understanding, although not one that could be used to transport energy, matter or any information instantiated in energy or matter.

SPACETIME

In considering spacetime some modern researchers are keen to stress the existence of spacetime as a reality rather than an abstraction, thus establishing it as something that can interact with the quanta and influence the development of the universe. Below, we look at the arguments of several researchers. They do not agree with one another, and it is beyond our scope here to discuss the merits of the different approaches. Rather there is an overall impression of spacetime as something which exists, has a discrete rather than continuous structure, and thus may contain coding, pattern or even decision making processes that influence the rest of the universe.

SPACETIME—CONTINUOUS OR DISCRETE

A century ago, Einstein showed that spacetime was not a fixed absolute background, or a rigid theatre within which life is acted out. Instead, it is conceived of as dynamic in response to changes in matter. In particular, general relativity described the functioning of gravity as the curvature of spacetime in response to massive objects.

In contrast to the other forces of nature, which have been shown to be intermediated by force carrying particles, the intermediation of the gravitational force or quantum gravity remains elusive. To understand quantum gravity would involve removing the conflicts between relativity and quantum theory.

Relativity describes gravity and the behaviour of the universe on large scales, while quantum theory describes the behaviour of particles at a scale where gravity could be ignored. However, although both have been exhaustively tested, the two theories are not compatible, with several conflicting features. The smooth continuous curvature of spacetime in relativity conflicts with the discreteness of the quanta. The dynamic nature of spacetime in general relativity contrasts with the fixed spacetime background in quantum theory. Finally, the determinism of relativity contrasts with the probabilistic and random nature of quantum theory.

In mathematical terms, equations that try to link the two theories end in infinities, indicating that something is wrong. Further problems arise in situations such as black holes and the Big Bang where matter is compressed down to a scale that is normally described by the quantum. General relativity implies that matter would be infinitely compressed at the singularity of a black hole, but quantum theory does not allow infinite compression.

LOOP QUANTUM GRAVITY

One attempt to resolve the incompatibility of relativity and quantum theory is loop quantum gravity (LQG). This approaches the nature of spacetime from the angle of quarks, the fundamental particles making up the protons and neutrons of the atomic nucleus. The force that binds the quarks together is known as the colour force and has field lines connecting charges on the quarks analogous to the electromagnetic field, in which field lines connect electrically charged particles.

THE QUARKS

However, the action of the colour force is the reverse of the electromagnetic force that falls off with distance. When quarks are close together there is little constraint on their movement, and the force between them seems quite weak. However, if the quarks are pulled apart the force holding them together increases, until it reaches a constant value that does not fall off with distance. This force which is weak at close quarters and strong at a distance is analogous to a an elastic band. Of course, this cannot be understood literally as a piece of material. We know from elsewhere in quantum theory that the forces of nature are quantised, with photons conveying the electromagnetic force. In this case, the connection between the quarks is suggested to be quantised, and is viewed as comparable to the quantised magnetic flux in superconductors that allow the dissipationless transfer of energy.

SPACE AND ENERGY TRANSMISSION

One possibility following from this is that space has some of the energy transmission properties of a superconductor. This is plausible in relation to the ideas of space/the quantum vacuum as being full of oscillating particles. The vacuum fluctuations are in this view seen as the transmitters of force. An alternative view is that the strings binding the quarks are themselves the fundamental entities and have no requirement for a field-type theory. A further view favours a kind of duality in which the strings and the field are dual aspects of an underlying reality. The idea of a non-continuous structure for spacetime helps to deal with the incompatibility of relativity and spacetime.

It is claimed that the most successful approaches to quantum gravity involve three basic ideas, *(1.)* that space is an emergent property, *(2.)* that it is based on something that is discrete rather than continuous, and *(3.)* that causality is fundamental to this. Loop quantum gravity is based on the idea of describing a field in terms of its field lines. The field lines are viewed as describing the geometry of space. Areas and volumes come in discrete units. There is a suggestion that knots and links in this network code for particles. Different ways of knotting or braiding the network are suggested to be different elementary particles. When coupled to the standard model of particle physics there are no infinities in the system. When applied to black holes and the Big Bang there is a suggestion that singularities are eliminated.

THE GEOMETRY OF SPACETIME

Loop quantum gravity emerged out of this thinking, with the geometry of spacetime expressed in terms of loops. These are the loops of colour force flux, but instead of being related to any fixed background, it is their interrelation that defines space. The theory depends on the particular connections between the loops. The theory allows the area of any surface to come not as any value but only in discrete multiples of units. The smallest units is about the Planck area, which is the square of the Planck length. Surfaces are made of discrete parts each of which comprises a finite amount and similarly for volumes. The geometry of spacetime is not fixed, but changes as a consequence of the movement of matter. Geometry here describes the relationships of lines or edges, areas and volumes with respect to space. The laws of physics govern how the geometry of spacetime evolves, but spacetime emerges from the laws rather than providing a stage for them to act on.

With loop quantum gravity, a position in space can only be defined in relation to other objects in space. Following this line of reasoning, it is the objects that create space. If there were no objects or only one object in the universe, there would be nothing that could be identified as space. Similarly, the motion of objects in space can be defined only in relation to other objects. The geometry of space, or the measurement of its areas and volumes, is seen as changing, when the position of objects alters relative to one another. Physicists refer to this approach as 'background independence', with objects evolving and creating their own spacetime, rather than operating in relation to a fixed background spacetime.

Most of the information needed to construct the geometry of spacetime comprises information about its causal structure. The fact that the universe is a causal structure means that even terms such as 'things' or 'objects' or 'objects in space' are not strictly correct, because causal structure means that things or objects are constantly developing, so that they are really processes rather than things, and as such causal structures that are creating spacetime geometry through their dynamic change.

PENROSE SPIN NETWORK

It was found that the discrete units of loop quantum gravity related to the spin network theory developed by Roger Penrose a generation earlier. Penrose had also considered that space was purely relational. The spin network was the version of quantum geometry that Penrose came up with. The spin network is a graph labelled with integers, with the spins that particles have in quantum theory. The spin networks provide a possible quantum state for the geometry of space. The edges of the network correspond to units of area. The nodes, where edges of the spin network meet, correspond to units of volume.

Further study suggested that the spin network picture follows from combining quantum theory with relativity. The spin networks are not set in space, they generate space, with relationships in space determined by how the edges come together at the nodes. The spin networks can evolve in time in response to changes. Each event in spacetime is seen as a change in the quantum geometry of space. The causal evolution of the spin networks can be described by the development of light cones through time. In Penrose's approach to conscious theory, consciousness and understanding is seen as being embedded in the geometry of spacetime as described in spin networks.

BLACK HOLES AND HIDDEN REGIONS

The extreme conditions of black holes are viewed as a way of examining the physics necessary to an understanding of quantum gravity, and may point ultimately to a connection between randomness and entanglement. A black hole has a horizon from within which even light cannot escape, because of the strength of the hole's gravitational force. This creates a region that is hidden from observers anywhere outside of the horizon of the black hole. The entropy of the horizon of a black hole is given as a quarter of the area of the horizon divided by h bar times the gravitational constant. Perhaps more helpfully, the horizon can be conceived as a computer screen with one pixel for every four Planck areas. The amount of information hidden in the black hole is equivalent to the number of notional pixels.

Quite apart from conditions around black holes, all observers have hidden regions, which are comprised of all those regions of the universe from which they will never receive light signals, because of their distance and the continuing expansion of the universe. The boundary between the part of the universe an observer can see and the part they cannot see is called an horizon.

ACCELERATING SPACESHIPS SEE HOT PARTICLES

In relation to this, the situation of an observer in a spaceship accelerating towards, but not reaching, the speed of light can be envisaged. It is emphasised that as that as long as the ship continues to accelerate, there will be a region behind it from which light does not catch up with the ship, and that this will constitute a hidden region for an observer on the ship. Working from the equivalence of gravity and acceleration, researchers Paul Davies and Bill Unruh also think that if an observer near a black hole saw heat radiation coming from the black hole, it means that an observer accelerating through the quantum vacuum would see heat radiation coming from in front of them. The significance of this is that from the point of view of an accelerated observer, the quantum vacuum is a real thing capable of having an effect.

This relates to uncertainty principle, a key concept within quantum theory, which prevents quantum particles from having a precisely defined position and momentum at the same time. The consequence of this is that even when a system is cooled to a point at which it has no energy, it will still have an intrinsic random motion, because if this ceased its position and momentum would both be precisely defined. This is known as zero point energy. Because of the lack of ordinary energy in the system, detectors do not register this motion. However, if the detector is accelerating, as would be the case of detectors placed on our observer's spaceship, the accelerated ship/detector is itself a source of energy that allows the zero point energy to be detected.

Uncertainty principle does not only apply to position and momentum, but also to the electric and magnetic fields that permeate space. One cannot simultaneously know the precise position of both the electric and the magnetic field in a particular region of space. Even when a region is cooled so as to contain zero energy, there will be randomly fluctuating electric and magnetic fields, referred to as quantum fluctuations of the vacuum. These, however, would also be detected by the accelerating detector on our observer's spaceship.

EXPERIMENTAL FOR ENERGY FROM EMPTY SPACE

A recent experiment by Chris Wilson et al at Chalmers University of Technology in Gothenburg serves to substantiate the prediction that energy in the form of photons could be derived from empty space. The prediction was based on uncertainty principle, which does not permit permanent zero energy, but specifies a fluctuation between zero and a small amount of energy. This comes in the form of virtual photons that jump in and out of existence, but can become real photons if they absorb energy.

The existence of these photons had already been inferred from the Casimir effect. In this, with metal plates very close to one another, longer wave length photons are excluded from the space between the plates, and as a result there is an inward pressure on the plates indicating the energy of the vacuum.

The Wilson experiment has gone beyond inference to the actual production of real photons from the vacuum. To achieve this, Wilson et al used a superconducting electrical circuit with an oscillator, which resulted in alterations in the distance an electron had to travel through the circuit. The alteration meant that the electron was doing the equivalent of travelling at a quarter of the speed of light. This proved sufficient for the kinetic energy of the electron to turn some of the virtual photons into real photons.

This is important in terms of how we conceive of spacetime, indicating that spacetime is a reality rather than an abstraction, and also that it is a reality in terms of discrete elements such as the virtual photons discussed here. This may make it more plausible to think in terms of spacetime having a fundamental measurement or geometry that can be related to a fundamental property of consciousness.

RANDOMNESS AND QUANTUM ENTANGLEMENT

It is possible to go a step further, and to try to explain where the randomness in the fluctuations of the electric and magnetic fields comes from. Going back to the example of the spaceship accelerating towards the speed of light, it is claimed that the photons that constitute the electric and magnetic fields are non-locally correlated, with each photon detected by the ship non-locally correlated with a photon in the ship's hidden region. The observed randomness is a measure of the observer's lack of information about the hidden region. The entropy represented by this randomness is related to the size of the hidden region. In turns out that the entropy of the particles detected by the ship is proportional to the area of the horizon of the ship's hidden region. This is referred to as Bekenstein's law, after the physicist of that name, stating that with the horizon of a hidden region there is an associated entropy that indicates the amount of information hidden by the region. Such things or processes are viewed as being finite in number, and therefore discrete from one another. Given that it is these events and processes that create space, it is therefore also possible to view spacetime as discrete. By contrast, the smooth continuous space implied by general relativity would require an infinite number of relationships. The discreteness of space is one of the few areas in speculative physics where there appears to be something of a consensus among physicists.

STRING THEORY

In the more popular rival to loop quantum gravity, string theory, the quanta are viewed as one-dimensional strings extending into higher dimensions, beyond the normal four dimensions. The extra dimensions are usually deemed to have been rolled up very small in the Big Bang, which accounts for them never having been detected. The manner in which the strings vibrate determines the nature of the particle involved. The analogy is that of the strings of a violin, where the vibration of the string determines the nature of the note. This has the advantage of being described by mathematics that would allow quantum theory and relativity to be compatible. It can also be speculated that spacetime as structured in string theory could support a discrete web or pattern that undergirded conscious experience and understanding, but this has not been extensively developed.

SPACETIME AS A FUNDAMENTAL

Other views favour the ideas of spacetime and the energy it contains as fundamental, while quantum particles are suggested to be less fundamental, existing as distortions or disturbances of the underlying spacetime.

Inertia is the built in resistance of objects to being moved if they are stationary, or having their motion changed if they are already moving. This kind of inertial mass is the most familiar form of mass. The associated concept of weight represents the force of gravity acting on the mass, and for this reason weight varies according to the local strength of the gravitational field. This is referred to as gravitational mass, as opposed to the constant of inertial mass.

HIGGS FIELD

Anything said in this area as of early 2012 is subject to current experiments at the Large Hadron Collider. One possible mechanism for endowing quanta with mass is the proposed Higgs field. The Higgs field is suggested to provide the 'rest mass' that is intrinsic to the particle rather than any mass associated with the energy of its movement. Fields such as the electromagnetic field and the Higgs field are here viewed as being fundamental, with quantum particles being less fundamental, because they are just local excitations of a field.

THE QUANTUM VACUUM

Still others think that mass comes from interaction between a quantum particle and the quantum vacuum, as the particle moves through the vacuum. The fundamental particles are seen as localised knots in the quantum fields. In this idea, quantum behaviour is traced back to the oscillation of photons jumping in and out of existence in the quantum vacuum.

This idea was developed in relation to Hawking radiation. Hawking proposed that the strong gravity near a black hole distorts the quantum vacuum so that virtual photons that normally pop in and out of existence here receive enough energy to become permanent particles. It is suggested that these permanent photons would to an external observer look like the radiation from a hot furnace, or the hot particles seen by the observer in the spacecraft discussed above.

The electric and magnetic fields flowing through space can be argued to constantly oscillate, as a function of the uncertainty of their position and momentum. The name 'zero-point field' refers to the fact that this is the lowest possible energy state that persists even when the heat/movement of molecules has ceased. Because electromagnetic radiation permeates the whole of space, this adds up to an enormous amount of energy. It is argued that there is no such thing in the universe as a void, and that this lowest energy state is still full of this zero point energy. This quantum vacuum is viewed as a sea of energy fluctuations and force perturbations jumping in and out of existence.

The zero point energy can be treated as a real thing, and concentrates attention on what effect this has. The existence of the zero point energy has long been demonstrated by the Casimir force. At distances smaller than a millimetre metal can be forced together, because long-wave radiation is suppressed between the plates, so more pressure is exerted on the metal sheets from outside than inside. The nearer the plates are brought together, the more radiation is excluded and the greater the external pressure.

MASS AND VACUUM

The assumption since Newton has been that the mass of an object, which is in effect a measure of its inertia, was an innate property of the object itself. However, this has been recently challenged by an opposite proposal that the inertial resistance to acceleration came not from the object itself, but instead from a contrary force exerted by the surrounding zero-point field. One suggestion is that the oscillation of the virtual particles of the vacuum interact with objects so as to produce inertial mass. Photons are seen as being exchanged between the virtual particles of the quantum vacuum and the quarks and electrons that are most fundamental in matter. This accords with the idea that inertial force comes from outside the body, from the quantum vacuum and from the interaction between the particles of matter and the virtual particles of the quantum vacuum.

In this approach that the fundamental thing is not mass, but the quantum vacuum. In this view Higgs field is relegated to producing rest mass, while inertial mass comes from the vacuum. Photons can be exchanged between the quantum vacuum and the quarks and electrons that make up matter. Although an electron is regarded as a point particle, it behaves as if it had a certain size, and this is viewed as an oscillation that reflects the oscillation of the quantum vacuum around it.

INERTIAL AND GRAVITATIONAL MASS

It is further suggested that inertial and gravitational mass share a common origin, which is that they both arise from the interaction of electron charges with the quantum vacuum. With this concept the electric charge in matter distorts the quantum vacuum in their vicinity, attracting or repelling virtual particles with the same or opposite charges. This distortion interacts with the charges in other matter creating a force of attraction between the two pieces of matter. One bit of mass only pulls on another via the quantum vacuum. The bending of light that is seen as a proof of the warping of space in general relativity is here explained in terms of a distortion of the quantum vacuum. Acceleration through the quantum vacuum results in resistance from the vacuum and this is seen as explaining inertia. According to the theory of general relativity spacetime is warped by energy, with mass being categorised as a form of energy.

In the quantum theory approach to this concept virtual photons that jump in and out of existence in the vacuum warp spacetime around themselves. The source of the energy that warps space in general relativity is the energy density of space or the amount of energy in a unit volume of space.

SPACETIME AND CONSCIOUSNESS

What is the significance of all this for consciousness studies? 'Fundamentalist' theories try to explain consciousness in terms of fundamental quantum features, which ultimately involves the nature of the quantum vacuum/spacetime. An understanding of this therefore becomes central to an understanding of the physical basis of consciousness. If spacetime/the quantum vacuum has discrete structure, as these proposals discussed above suggest, it becomes the more plausible that it could provide a network, pattern or code that underlies consciousness.

SECTION 3

QUANTUM BIOLOGY

In this section we introduce recent research indicating the existence of quantum coherence in organic matter and look at the possible implications of this for our understanding of neurons. First, however we need to make an excursion into the physics underlying some long-established organic chemistry, which is relevant to the systems we discuss in this section. Once again no apologies for bringing you to a hard place. Those who are familiar with this basic material can move forward a few pages to the 'Quantum coherence and entanglement in biological systems' part of Section 3. That section may appear repetitive in places with mention of scientific papers that cover overlapping ground. I have retained this because I wish to demonstrate the depth of recent research in this area, which could be at risk of glibly being dismissed as pseudoscience.

π ELECTRONS

We start here by discussing the role of electrons around atoms. The overlap of the atomic orbitals forms bonds between atoms, and thus creates molecules, and also determines the shape of a molecule. The same atoms held in a different shape can result in a different compound. The term 'n' is used to describe the energy level of each orbital. Each value of 'n' can represent a group of orbitals at different energy levels known as a shell. The first shell, n = 1, can only contain one orbital, the second shell, n = 2, can contain two orbitals, the third shell, n = 3, can contain three orbitals and so on.

ANGULAR MOMENTUM

Another quantum number 'L' relates to the angular momentum of an electron in an orbital. The value of 'L' is at least one less than the value of 'n'. The values for 'L' are conventionally given by letters. For our purposes here we need only deal with the values of 0 and 1, which are labelled 's' and 'p'. So an electron can be labelled 2s, denoting an orbital energy of 2 and an angular momentum of 0, or it can be labelled 2p with an orbital energy of 2 and an angular momentum of 1.

ELECTRON WAVE FUNCTION

The electron orbital is viewed as being a wave function. With a wave function, the wavelength or its reciprocal, frequency, is related to the energy level of the individual quanta, but the amplitude (the height of the wave) squared is the strength of the signal, or in other words the number of quanta involved. With a photon, the quanta of light, frequency determines the colour of visible light, but the square of the amplitude, signifying the number of quanta, determines the brightness.

SPHERES AND LOBES

It is possible to chart the probability of an electron being present at a particular point in space, and this can be referred to as a density plot. For an 's' orbital (see angular momentum para. above) the density plot is spherical, but with 'p' electrons, the shape of the density plot is two lobes with a nodal area in between, where there is no electron density. The wave functions of these two lobes are out-of-phase.

A further quantum number m_L relates to the spatial orientation of the orbital angular momentum. The 's' orbitals have a 0 because a sphere does not have an orientation in space. For 'p' orbitals there are three possibilities of -1, 0 and +1 that can be related to the mutually perpendicular x, y, and z axes in geometry, and are written as p_x, p_y and p_z.

STRUCTURE OF AN ATOM

The structure of an atom involves having electrons in the lowest energy orbital and working up from there. Hydrogen has one electron located in the lowest energy orbital, and helium has two electrons placed in this orbital. Two electrons render an orbital full. An orbital can be full (2 electrons), half-full (one electron) or empty. With lithium which has three electrons, the third electron has to be located in a second orbital. With carbon there are six electrons, with two in the 'n' = 1, first shell. In the second, 'n' = 2 shell, there is one full orbital with two 's' electrons and two half-full orbitals each with one 'p' electron.

STRUCTURE OF MOLECULES

The structure of the individual atom is also the basis for the structure of molecules. Atomic orbitals are wave functions, and the orbital wave functions of different atoms are like waves, in that if they are in phase, their amplitudes are added together. When this happens, the increased amplitude of the wave function works against the mutual repulsion of the positively charged atomic nuclei of different atoms, and works to bond the atoms together. This is referred to as a bonding molecular orbital.

When the orbitals are out-of-phase, they are on the far sides of the atomic nuclei, which continue to repel one another due to like positive electric charges, and this arrangement is known as the anti-bonding molecular orbital. Collectively the two types of molecular orbital are referred to as MOs. The antibonding MOs usually have higher energy than the bonding MOs. Energy applied to an atom can promote a low-energy bonding orbital to a higher-energy anti-bonding orbital, and this process can break the bond between two atoms. When 's' orbitals combine, the MOs are symmetrical, and this type of orbital overlap has sigma (σ) symmetry, and is described as a sigma (σ) bond.

When there is a combination of 'p' orbitals, there is a possibility of three different 'p' orbitals on axes that are perpendicular to one another. One of these can overlap end-on with an orbital in another atom, and these two orbitals are described as $2p\sigma$ and $2p\sigma^*$. Two other orbitals can overlap with those on other atoms side-on, and will not be symmetrical about the nuclear axis. These are described as π orbitals and form π bonds.

In discussing bonding, only the electrons in the outermost shell of the atoms are usually relevant. For example, in a nitrogen molecule formed by the bonding of two nitrogen atoms, only the electrons in the second, 'n' = 2, shell are involved in bonding. The nitrogen atom has seven electrons, so there are fourteen on the two atoms that bond to form a nitrogen molecule. Two electrons in the inner shell of each atom are not involved, leaving five on each atom and ten altogether in the second shells. The 2s electrons on each atom cancel out, and are described as lone pairs.

The bonding work thus devolves on three electrons in each atom, or six in the whole molecule. These form one σ bond and two π bonds. This is described as a triple-bonded structure. Orbitals overlap better when they are in the same shell of their respective atoms. So electrons in the second shell will overlap more readily with second shell electrons in other atoms than with third or fourth shell electrons in these other atoms. Further to that 'p' electrons must have the right orientation and p_x electrons can only interact with other p_x electrons and so on, because the x, y and z electrons are perpendicular or orthogonal to one another.

Molecular bonding also applies to molecules that are formed out of different types of atoms, as distinct from molecules formed from atoms of the same element such as the nitrogen molecule discussed above. If the atomic orbitals of different atoms are very different, they cannot combine, and the atom cannot form covalent bonds (sharing the electron between two atoms). Instead an electron can transfer from one atom to another, transforming both the atoms into ion, and the second atom into a positive ion, with the molecule now held together by the attraction between the oppositely charged ions. This is known as ionic bonding. Covalent bonds with overlapping orbitals can only be formed when the difference in energy is not too great.

HYBRIDISATION

Hybridisation is an important factor in the formation of molecular bonds. The 's' and 'p' orbitals are those most important for organic chemistry, and for the bonding of atoms such carbon, oxygen, nitrogen, sulphur and phosphorous. Hybridised orbitals are viewed as 's' and 'p' orbitals superimposed on one another.

In its ground state, the carbon atom has two electrons in the first shell, and this is not normally involved in bonding. In its second and outer shell, it has two 's' electrons filling an orbital, and two 'p' electrons, one p_x and one p_y, each in a half-filled orbital. If the carbon atom is excited, say by the positive charge attraction of the nucleus of a nearby hydrogen atom, an 's' electron in the outer shell can be excited into a 'p' orbital, so that the outer shell now has one 's' electron and three 'p' electrons, one each in an x, y and z orientation. The four outer shell electrons are now deemed to be not distinct 's' and 'p' electrons but four 'sp' electrons, here described as sp^3, because the configuration is one quarter 's' electron and three-quarters 'p' electrons. The arrangement allows the formation of four σ covalent bonds. Carbon atoms can also use sp^2 hybridisation where one 's' electron and two 'p' electrons in the outer shell are hybridised. There is also 'sp' hybridisation where the 's' orbital mixes with just one of the 'p' orbitals.

With the C=O double bond, the two atoms in the double bond are sp^2 hybridised. The carbon atom uses all three orbitals in the sp^2 arrangement to form σ bonds with other orbitals, but the oxygen atoms use only one of these. In addition a 'p' electron from each atom forms a π bond.

DELOCALISATION AND CONJUGATION

The joining together or conjugation of double bonds is important for organic structures. π bonds can form into a framework over a large number of atoms, and are seen to account for the stability of some compounds. The structure of benzene is relevant in this respect. Benzene is based on a ring of six carbon atoms. The carbon atoms are sp^2 hybridised, leaving one 'p' electron per carbon atom free, or six electrons altogether. These six electrons are spread equally over the six carbon atoms of the ring. These are π bonds delocalised over all six atoms in the carbon ring, rather than being localised in particular double bonds.

Delocalisation emphasises the spatial spread of the electron waves, and occurs over the whole of the conjugated system. This is sometimes referred to as resonance. Sequences of double and single bonds also occur as chains rather than rings. Conjugation refers to the sequence of single and double bonds that form either a ring or a chain. Double bonds between carbon and oxygen can be conjugated in the same way as double bonds between carbon atoms. Conjugation involves there being only one single bond between each double bond. Two double bonds together also do not permit conjugation. These 'rules' relate to the need to have 'p' orbitals available to delocalise over the system.

In both rings and chains every carbon atom is sp^2 hybridised leaving a third 'p' electron to overlap with its neighbours, and form an uninterrupted chain. The double bonds that are conjugated with single bonds are seen to have different properties from double bonds not arranged in this way. Here again conjugation leads to a significantly different chemical behaviour.

Chlorophyll, the pigment molecule in plants, is a good example of a conjugated ring of single and double bonds, and the colour of all pigments and dyes depends on conjugation. The colour involved depends on the length of the conjugated chain. Each bond increases the wavelength of the light absorbed. With less than eight bonds light is absorbed in the ultra-violet.

The colours of objects and materials around us are a function of the interaction of light with pigments. For instance, pigments are characterised by having a large number of double bonds between atoms. The pigment, lycopene, responsible for the red in tomatoes and some berries, comprises a long chain of alternating double and single bonds, allowing the molecule to form π bonds. An extensive network of π bonds across a large number of atoms is involved in the chemistry of many compounds. It is responsible for the high degree of stability in aromatic compounds such as benzene.

The compound ethylene (CH_2=CH_2) has all its atoms in the same plane, and is therefore described as planar. In this molecule, the two carbon atoms are joined by a double bond. Hybridisation involves mixing the 2s orbital on each carbon atom with two out of the three 'p' orbital on each carbon atom to give three sp_2 orbitals. The third 'p' orbital on each atom overlaps with the 'p' orbital of the other atom to form a π bond. The 'p' orbitals of the two atoms also have to be parallel to one another in order to form a π bond. This prevents the rotation of the double bond between the carbon atoms. However, sufficient energy, such as that of ultra violet light, can break the π bond, and thus allow the double bond to rotate.

AROMATIC MOLECULES

An important feature of benzene is the ability to preserve its ring structure through a variety of chemical reactions. Benzene and other compounds that have this property are termed aromatic. In looking at these structures, the important feature is not the number of conjugated atoms, but the number of electrons involved in the π system The six π electrons of benzene leave all its molecular orbitals fully occupied in a closed shell, and account for its stability. A closed shell of electrons in bonding orbitals is a definition of aromacity.

In benzene, the lowest energy 'p' orbitals comprise electron density above and below the plane of the molecule. These electron orbitals are spread over, delocalised over or conjugated over all six carbon molecules in the benzene ring. The delocalised 'p' orbitals can themselves be thought of as a ring. Expressed another way, this type of delocalisation is an uninterrupted sequence of double and single bonds, and it is this which is described as conjugation. The properties of this type of system are seen to be different from its component parts.

Benzene has six π electrons, and in consequence all its bonding orbitals are full, giving the molecule a closed structure, which is often not the case for quite similar molecules with a lot of double bonds. This is referred to as a molecule being aromatic. The general rule is that there has to be a low energy bonding orbital with the 'p' orbitals in-phase. There is a closed shell giving greater stability in aromatic systems, where there are two 'p' orbitals forming a π bond and four other electrons.

CARBON AND OXYGEN BONDS

It is not essential in these systems to have carbon-to-carbon bonds. Carbon and oxygen also often form double bonds, separated by just one single bond. Here to the behaviour of the double-bonded system is quite different from the behaviour of the component parts. These structures are special in the sense of only arising where there are 'p' orbitals on different atoms available to overlap with one another. In many other molecules, there is a similarity in terms of a large number of double bonds, but they are insulated from one another by the lack of 'p' orbitals available to overlap with one another.

AMIDE GROUPS, AMINO ACIDS AND PROTEIN

The amide group is crucial to protein, and therefore to living systems as a whole, in that it forms the links between amino acid molecules that in turn make up protein, the basic building blocks of life. The amino group on one amino acid molecule combines with the carboxylic group on another amino acid molecule to create an amide group. When a chain of this kind forms it is a peptide or polypeptide, and longer chains are classed as proteins. Conjugation arises from the bonding of a lone pair of 'p' orbitals, and this is vital in stabilising the link between the amino acids, and making it relatively difficult to disrupt the amino acid chains that make up protein.

QUANTUM COHERENCE AND ENTANGLEMENT IN BIOLOGICAL SYSTEMS

The key argument against quantum states having a practical role in neural processing is that in the conditions of the brain quantum decoherence would happen too rapidly for the states to be relevant to neural processes. This view was crystallised by Max Tegmark's paper in 2000 (9.) published in the prestigious journal, Physical Review E. The paper itself was not remarkable. For reasons that have never been properly explained, it used a model of quantum processing that has never been proposed elsewhere, and it failed to discuss or even mention arguments for the shielding of quantum processing in the brain. Nevertheless, it succeeded in confirming in a prestigious way the views of the numerous opponents of quantum consciousness.

The situation remained like that between 2000 and 2007, after which the debate over quantum states in biological systems was moved to a new stage. The publication of Engel et al's paper in 'Nature' in 2007 demonstrated that contrary to the main thrust of Tegmark and those that relied on him, quantum coherence has a functional role in the transfer of energy within organisms, in this case photosynthetic organisms. This moved the discussion of what sort of coherent biological features could support consciousness on from a phase of pure theorising, to a phase, in which ideas can be related to features that have been shown to exist in biological matter.

THE ENGEL STUDY

The Engel et al paper studied photosynthesis in green sulphur bacteria. The photosynthetic complexes in the bacteria are tuned to capturing light and transmitting its energy to long-term storage areas. It should be stressed that in this system, photons (the light quanta) only provide the initial excitation, and the coherence and entanglement discussed here involves electrons within biological systems.

The Engel study documented the dependence of energy transport on the spatially extended properties of the wave function of the photosynthetic complexes. In particular, the timescale of the quantum coherence observed was much longer than would classically be predicted for a biological environment, with a duration of at least 660 femtoseconds (femtosecond=10^{-15} seconds), nearly three times as long as the classically predicted times of 250 femtoseconds. In the latter case, rapid destruction of coherence would prevent it from influencing the system. The wavelike process noted by Engel was suggested to account for the efficiency of the system, at 98% compared to the 60-70% predicted for a classical system.

LIMITED DEPHASING

Another researcher in this area, Martin Plenio, argues that where temperatures are relatively high, there is likely to be some dephasing of the quanta, but contrary to the popular view that this would be the end of quantum processing, the efficiency of energy transportation could actually be enhanced by this limited dephasing. Referring to a quantum experiment with beam splitters and detectors, he suggests that partial dephasing might actually allow the wider and therefore more efficient exploration of the system.

CHENG & FLEMING:—THE PROTEIN ENVIRONMENT.

In a paper by Cheng & Fleming published in 'Science' (11.) a study of long-lived quantum coherence in photosynthetic bacteria, demonstrates strong correlations between chromophore molecules. One experiment looked at two chromophore molecules. The system provided near unity efficiency of energy transfer, and also demonstrates energy transfer between the chromophores.

The experiment also shows that the time for dephasing of these molecules is substantially longer than would have been classically estimated. The traditional approach in particular ignored the coherence between donor and acceptor states. The adaptive advantages of this lie in the efficiency of the search for the electron donor. The longer time to dephasing of one as compared to the other of the experimental chromophores was taken to indicate a strong correlation of the energy fluctuations of the two molecules. This meant that the two molecules were embedded in the same protein environment.

Another study by Fleming et al that also observed long-lasting coherence in a photosynthetic organism indicated that this could be explained by correlations between protein motions that modulate the transition energies of neighbouring chromophores. This suggests that protein environments works to preserve electronic coherence in photosynthetic complexes, and thus optimise excitatory energy transfer.

CHAINS OF POLYMERS

Elizabetta Collini and Gregory Scholes conducted an experiment also reported in 'Science' (12.) that observed quantum coherence dynamics in relation to electronic energy transfer. The experiment examined polymer samples with different chain conformations at room temperature, and recorded intrachain, but not interchain, coherent electronic energy transfer.

It is pointed out that natural photosynthetic proteins and artificial polymers organise light absorbing molecules (chromophores) to channel photon energy. The excitation energy from the absorbed light can be shared quantum mechanically among the chromophores. Where this happens, electronic coupling predominates over the tendency towards quantum decoherence, (loss of coherence due to interaction with the environment), and is viewed as comprising a standing wave connecting donor and acceptor paths, with the evolution of the system entangled in a single quantum state. Within chains of polymers there can be conformational subunits 2 to 12 repeat units long, which are the primary absorbing units or chromophores. Neighbouring chromophores along the backbone of a polymer have quite a strong electronic coupling, and electronic transfer between these is coherent at room temperature.

QUANTUM ENTANGLEMENT CONSIDERED—SAROVAR ET AL (2009)

In a 2009 paper, Sarovar et al (13.) examined the subject of possible quantum entanglement, as distinct from quantum coherence, in photosynthetic complexes. The paper starts by discussing quantum coherence between the spatially separated chromophore molecules found in these systems. Modelling of the system showed that entanglement would rapidly decrease to zero, but then resurge after about 600 femtoseconds. Entanglement could in fact survive for considerably longer than coherence, with a duration of five picoseconds at 77K, falling to two picoseconds at room temperature. The entanglement examined here is the non-local correlation between the electronic states of spatially separated chromophores. Coherence is a necessary and sufficient state for entanglement to exist.

ISHIZAKI AND FLEMING (2009)

This paper (14.) developed an equation that allows modelling of the photosynthetic systems discussed above. Where this deals with the sites to be excited by the light energy, the initial entanglement rapidly decreases to zero, but then increases again after about 600 femtoseconds. This is thought to be a function of the entanglement of the initial sites being transported and localised at other sites, but remaining coherent at these other sites, from which further entanglement can subsequently resurge.

Other studies appear to confirm the existence of picosecond timescales for entanglement in chromophores. It is not clear to the authors that entanglement is actually functional in chromophores. Coherence appears to be sufficient for very efficient transport of energy, and entanglement may be only a by-product of coherence. This looks to remain an area of scientific debate.

Earlier studies such as Engel's were performed at low temperatures, whereas quantum coherence becomes more fragile at higher temperatures, because of the higher amplitude of environmental fluctuations. In the Ishizaki and Fleming paper, the equation supplied by the authors suggest that coherence could persist for several hundred femtoseconds even at physiological temperatures of 300 Kelvin.

This study deals with the Fenna-Matthews-Olson (FMO) pigment-protein complex found in low light-adapted green sulphur bacteria. The FMO is situated between the chlorosome antenna and the reaction centre, and its function is to transport energy harvested from sunlight by the antenna to the reaction centre. The FMO complex is a trimer of identical sub-units, each comprised of seven bacteriochlorphyl (BChl) molecules. This structure has been extensively studied.

Each unit of the FMO comprises 7 BChl molecules. BChl 1 and 6 are orientated towards the chlorosome antenna, and are the initially excited pigment, and BChl 3 and 4 are orientated towards the reaction centre. Even at the physiological temperatures, quantum coherence can be observed for up to 350 femtoseconds in this structure. This suggests that long-lived electronic coherence is sustained among the BChls, even at physiological temperatures, and may play a role in the high efficiency of EET in photosynthetic proteins.

BChl 1 and 6 are seen as capturing and conveying onward the initial electronic energy excitation. Quantum coherence is suggested to allow rapid sampling of pathways to BChl 3 that connects to the reaction centre. If the process was entirely classical, trapping of energy in subsidiary minima would be inevitable, whereas quantum delocalisation can avoid such traps, and aid the capture of excitation by pigments BChl 3 and 4. BChl 6 is strongly coupled to BChl 5 and 7, which are in turn stongly coupled to BChl 4, ensuring transfer of excitation energy.

Delocalisation of energy over several of the molecules allows exploration of the lowest energy site in BChl 3. The study predicts that quantum coherence could be sustained for 350 femtoseconds, but if the calculation is adjusted for a possible longer phonon relaxation time, this could extend to 550 femtoseconds, still at physiological temperatures.

CIA ET AL (2008):—RESETTING ENTANGLEMENT

A 2008 paper from Cia et al (15.) also looked at the possibility of quantum entanglement in the type of system studied in the Engel paper. Cia takes the view that entanglement can exist in hot biological environments. Cia says traditional thinking on biological systems is based on the assumption of thermal equilibrium, whereas biological systems are far from thermal equilibrium. He points out that the conformation of protein involves interactions at the quantum level. These are usually treated classically, but Cia wonders whether a proper understanding of protein dynamics does not require quantum mechanics. It is said not to be clear, whether or not entanglement is generated during the motions of protein, but that it is possible that entanglement could have important implications for the functioning of protein.

The model studied by the Cia et al paper suggests that while a noisy environment, such as that found in biological matter, can destroy entanglement, it can also set up fresh entanglement. It is argued that entanglement can recur in the case of an oscillating molecule, in a way that would not be possible in the absence of this oscillation.

The molecule has to oscillate at a certain rate relative to the environment to become entangled. This process allows for entanglement to emerge, but this would normally also disappear quickly. Something extra is needed for entanglement to recur or persist. It is suggested here that the environment, which is normally viewed as the source of decoherence, can play a constructive role in resetting entanglement, when combined with classical molecules. Environmental noise in combination with molecular motion provides a reset mechanism for entanglement. According to the author's calculations entanglement can persistently recur in an oscillating molecule, even if the environment is too hot for static entanglement. The oscillation of the molecule combined with the noise of the environment may repeatedly reset entanglement.

THE FMO COMPLEX AND ENTANGLEMENT

A paper by K. Birgitta Whaley, Mohan Sarovar and Akihito Ishizaki (16.) published in 2010 discusses recent studies of photosynthetic light harvesting complexes. The studies are seen as having established the existence of quantum entanglement in biologically functional systems that are not in thermal equilibrium. However, this does not necessarily mean that entanglement has a biological function. The authors point out that the modern discussion of entanglement has moved and from simple arrangements of particles to entanglement in larger scale systems.

Measurements of excitonic energy transport in photosynthetic light harvesting complexes show evidence of quantum coherence in these systems. A particular focus of research has been the Fenna-Matthew-Olson (FMO) complex in green sulphur bacteria. The FMO serves to transport electronic energy from the light harvesting antenna to the photosynthetic reaction centre. Coherence is present here at up to 300K. The authors draw attention to the relationship between electronic excitations in the chromophores and those in the surrounding protein.

The electronic excitations in the chromophores are coupled to the vibrational modes of the surrounding protein scaffolding. One study shows a correlation between the extent of entanglement and the efficiency of the energy transport. The study went on to claim that efficient transport requires entanglement, although the authors of this paper query such a definite assertion.

The pigment-protein dynamics generates entanglement across the entire FMO complex in only 100 femtoseconds, but are followed by oscillations that damp out over several hundred femtoseconds, with a subsequent longer contribution continuing beyond that for up to about five picoseconds. This more persistent entanglement can be at between a third and a half of the initial value and 15% of the maximum possible value. Long-lived entanglement takes place between four or five of the existing seven chromophores. The most extended entanglement is between chromophores one and three, and these are also two of the most widely separated chromophores.

Studies also show that this entanglement is quite resistant to temperature increase, with only a 25% reduction when the temperature rises from 77K to 300K. Overall studies indicate long-lived entanglement of as much as five picoseconds between numbers of excitations on spatially separated pigment molecules. This is described here as long-lived coherence because energy transfer through the FMO complex is on a time span of a few picoseconds meaning that the up to five picoseconds of entanglement seen between the chromophores represents a functional timescale. However, the authors do not consider this by itself to be a conclusive argument for entanglement being functional in the FMO.

LIGHT-HARVESTING COMPLEX II (LNCII)

This paper also looks at light harvesting complex II (LHCII), which is also shown to have long-lived electronic coherence. LHCII is the most common light harvesting complex in plants. The system comprises three subunits each of which contains eight chlorophyll 'a' molecules and six chlorophyll 'b' molecules. A study by two of the authors (Ishizaki & Fleming, 2010) indicates that only one out of chlorophyll molecules would be initially excited by photons, and this molecule would then become entangled with other chlorophyll molecules. Entanglement decreases at first, but then persists at a significant proportion of the maximum possible value. This is also an important feature of the FMO complex.

In both these complexes entanglement is seen to be generated by the passage of electronic excitation through the light harvesting complexes, and to be distributed over a number of chromophores. Entanglement persists over a longer time and is more resistant to temperature increase than might have been previously expected. A functional biological role is suggested by the persistence of entanglement over the same timescale as the energy transfer within the light harvesting complexes.

Light harvesting complexes (LHCs) are densely packed molecular structures involved in the initial stages of photosynthesis. These complexes capture light, and the resulting excitation energy is transferred to reaction centres, where chemical reactions are initiated. LHCs are particularly efficient at transporting excitation energy in disordered environments. Simulations of the dynamics of particular LHCs predict that quantum entanglement will persist over observable timescales. Entanglement here would mean that there are non-local correlations between spatially separated molecules in the LHCs.

The molecules in the LHCs, referred to as chromophores, are close enough together for considerable dipole coupling leading to coherent interaction over observable timescales. The existence of coherence between molecules in these systems has been recognised for a decade or more. This condition is seen as the basis for entanglement. Coherence in this area, known as the site basis, is necessary and sufficient for entanglement, and any coherence in the area will lead to entanglement, and can be viewed in experiments as a signature of entanglement.

The authors base part of their study on the description of the dynamics of a molecule in a protein in an LHC. This model indicates the coupling of some pairs of molecules due to proximity and favourable dipole orientation, thus effectively forming dimers. The wave function of the system is delocalised across these dimers.

Using this equation, the interface of the LHC with light energy leads to a rapid increase in entanglement for a short time, followed by a decay punctuated by varying amounts of oscillation. The initial rapid increase reflects the coherent coupling of some parts of the

LHC system. This entanglement decreases again as the excitation comes into contact with other parts of the protein. Some of the entanglement seen is not between immediately neighbouring molecules, but between more distant parts of the LHC. Entanglement in LHC is estimated to continue until the excitation reaches the reaction centre. The authors view this as a remarkable conclusion, since it shows that entanglement between several particles can persist in a non-equilibrium condition, despite being in a decoherent environment.

ENTANGLEMENT AND EFFICIENCY

A paper by Francesca Fassioli and Alexandra Olaya-Castro (17.) suggests that electronic quantum coherence amongst distance donors could allow precise modulation of the light harvesting function. Photosynthesis is remarkable for the near 100% efficiency of energy transfer. The spatial arrangement of the pigment molecules and their electronic interaction is known to relate to this efficiency.

Recent experimental studies of photosynthetic protein have shown that it can sustain quantum coherence for longer than previously expected, and that this can happen at the normal temperature of biological processes. This has been taken to imply that quantum coherence may affect light harvesting processes. In photosynthesis, the energy of sunlight is transferred to a reaction centre with near 100% efficiency. The spatial arrangement of pigment molecules and their electronic interactions is known to be involved with this high efficiency. There is an implication that quantum coherence may affect the light harvesting process.

Some studies point to very efficient energy transport as the optimal result of the interplay of quantum coherent with decoherent mechanisms. Roles proposed for quantum coherence vary between avoidance of energy traps that are not at the overall lowest energy level, and actual searches for the overall lowest energy level. In this paper, it is suggested that the function of quantum coherence goes beyond efficiency of energy transport, and includes the modulation of the photosynthetic antennae complexes to deal with variations in the environment.

THE ROLE OF QUANTUM ENTANGLEMENT

There is some debate as to whether quantum entanglement, as distinct from coherence, plays a role in the functioning of the light-harvesting complexes, or is just a by-product of quantum states. The authors here argue that entanglement may be involved in the efficiency of the system, and they use the FMO protein in green sulphur bacteria as the basis of their study. They suggest that entanglement could play a role in light-harvesting by allowing precise control of the rate at which excitations are transferred to the reaction centre.

Long-range quantum correlations have been suggested to be important as a mechanism helping quantum coherence to survive at the high temperatures sustained in light harvesting antennae. This paper claims to show that in the FMO complex long-lived quantum coherence is spatially distributed in such a way that entanglement between pairs of molecules controls the efficiency profile needed to cope with variations in the environment. The ability to control energy transport under varying environmental conditions is seen as crucial for the robustness of photosynthetic systems. A mechanism involving both quantum coherence and entanglement might be effective in controlling the response to different light intensities.

ROOM TEMPERATURE: MOVING THE DEBATE FORWARD

A paper by Elizabetta Collini et al published in 'Nature' in 2010 (18.) moved the debate forward in an important way by demonstrating the existence of room-temperature quantum coherence in organic matter.

This paper describes X-ray crystallography studies of two types of marine cryptophyte algae that have long-lasting excitation oscillations and correlations and anti-correlations, symptomatic of quantum coherence even at ambient temperature. Distant molecules within the photosynthetic protein are thought to be connected to quantum coherence, and to produce efficient light-harvesting as a result. The cryptophytes can photosynthesise in low-light conditions suggesting a particularly efficient transfer of energy within protein. According to the classical theory, this would imply only small separation between chromophores, whereas the actual separation is unusually large.

In this study, performed at room temperature, the antenna protein received a laser pulse, resulting in a coherent superposition. The experimental data of the study shows that the superposition persists for 400 femtoseconds and over a distance of 2.5 nanometres. Quantum coherence occurs in a complex mix of quantum interference between electronic resonances, and decoherence is caused by interaction with the environment. The authors think that long-lived quantum coherence facilitates efficient energy transfer across protein units.

The authors remains uncertain, as to how quantum coherence can persist for hundreds of femtoseconds in biological matter. One suggestion is that the expected rate of decoherence is slowed by shared or correlated motions in the surrounding environment. Where light-harvesting chromophores are covalently bound to the protein backbone, it is suggested that this may strengthen correlated motions between the chromophores and the protein. Covalent binding to the protein backbone is speculated to make coherence longer lasting.

WIDESPREAD IN NATURE

In addition to the discovery of quantum coherence in biological systems at room temperature, studies now also show that coherence is present in multicellular green plants. Calhoun et al, 2009 (19.) studied this kind of organism. These two discoveries, coherence at room temperature and coherence in green plants have removed the initial possibility that coherence in photosynthetic organisms was an outlier confined to extreme conditions rather than something that was widespread in nature.

POSSIBLE EXPERIMENTAL SIMULATION OF ENTANGLEMENT IN PROTEIN:

In their paper 'Persistent dynamic entanglement from classical motion: How bio-molecular machines can generate non-trivial quantum states', Guerreschi, G., Cai, J., Popescu, S. & Briegel, H., from the Universities of Innsbruck, Ulm and Bristol (58.), discuss a model that studies the cyclic regeneration of quantum entanglement in hot systems. This looks to open the road to modelling or even experimental simulation that would constitute a possible test for/ falsification of non-trivial quantum states in proteins such as those found in neurons.

The paper refers to a simple mechanism by which a molecule forced out of thermal equilibrium by oscillations, can sustain quantum entanglement. This type of entanglement can survive intense noise, but cannot survive if the oscillation ceases. This is argued to be the basis for non-trivial quantum entanglement in biological matter.

The authors remark that this reverses the previous orthodoxy, which held that quantum effects could not exist in biological systems because of the amount of noise in these systems. They note that research in photosynthetic organisms have undermined this case in recent years. The existence of entanglement in a system is seen as greatly increasing information processing capacity, and this underlies the potential of quantum computing.

Thermal equilibrium: It is pointed out that the previous orthodoxy was based on the assumption of thermal equilibrium, whereas biological systems are open and driven systems far from thermal equilibrium. Such systems are suggested to be capable of quantum error correction that could sustain longer-lived quantum entanglement in biological systems.

In an earlier (2010) paper in Phys Rev E (1.) the authors presented a mechanism by which a molecule subjected to non-thermal equilibrium oscillations could sustain entanglement between two states. This could be maintained despite a level of environmental noise that would not allow entanglement to persist in the absence of non-equilibrium oscillations. Protein molecules, which undergo conformational changes are suggested as the sort of environment in which quantum entanglement of the type found in this model could arise.

In the first section of their paper, the authors look at the possibility of entanglement generated by molecular motion. A biomolecule undergoing conformational change can lead to an interaction between different sites of the molecule. The conformational changes of the molecule can force localised spins to come close or move apart. With the molecular configuration oscillating in a periodic way, cyclic regeneration of entanglement can be sustained over long periods of time, despite noise that would make static entanglement impossible. With thermal equilibrium, entanglement becomes impossible above a certain temperature. The authors, however, ask what happens when molecular motion is involved, and seek to demonstrate that entanglement can keep recurring in an oscillating molecule despite a hot environment.

The authors consider a simple process, with spins that are far apart and with an interaction that is weaker than the surrounding field. In this state, there will be no entanglement. When the spins approach one another entanglement can appear transiently on time scales shorter than that required for thermalisation. The molecule is seen as being kicked out of thermal equilibrium. The generation of entanglement depends on the rate of thermalisation not being too fast. The sustained recurrence of entanglement requires a persistent supply of free energy that can be produced by the conformational changes of the protein. In the author's model the background field predominates when the spins of the particles are widely separated, but when they are close together their interaction predominates. The authors assume that two spins start far apart and are in a state of thermal equilibrium. The spins oscillate, move closer together, are driven out of thermal equilibrium, and entanglement is generated. Environmental noise here drives a persistent and cyclic generation of new entanglement. The periodic oscillations are seen to keep molecules far away from thermal equilibrium, with the continuous change in the shape of the molecule preventing thermalisation.

The authors emphasise the constructive role played by thermalisation. In a hot thermal bath the first oscillation of the molecule is lost more quickly than in a cooler environment. However, the pumping of energy is seen to provide a reset mechanism. In discussing biological systems, the authors consider that chemical interactions would serve to keep the system out of equilibrium. But in gaps between chemical activity, equilibrium could return, and entanglement would therefore be transient.

In summary, the authors say that they have demonstrated that entanglement can recur even in a hot noisy environment. In biological systems this can be related to changes in the conformation of macromolecules. The authors say that this modelling is a route by which to search for the signatures of entanglement in biomolecular systems.

Experimental possibilities: They also think that existing technology could provide an experimental simulation of their model. From the distinct point of quantum consciousness studies, this could possibly amount to a test for/falsification of the hypothesis that non-trivial quantum states act within proteins, and thus test related theories of consciousness.

RELEVANCE TO OTHER ORGANISMS

The question arises as to whether quantum coherence and entanglement in plants has any relevance to animal life and in particularly to brains. A brief talk by Travis Craddock of the University of Alberta at a 2011 consciousness conference suggested that it could.

Craddock stressed that light absorbing chromophore molecules involved in light harvesting use dipoles to provide 99% efficiency in energy transfer from the light harvesting antennae to the reaction centre. The studies show that instead of quantum coherence being destroyed by the environment within the organism, a limited amount of noise in the environment acts to drive the system.

TRYPTOPHAN

Craddock indicates that any system of dipoles could work like this. He is particularly interested in the role of the amino acid, tryptophan. Similar models can be used for chromophores in photosynthetic systems and for tryptophan, an aromatic amino acid that is one of the 20 standard amino acids making up protein, including the microtubular protein, tubulin.

Tryptophan has eight molecules extending over the length of the tubulin protein dimer, and it possesses strong transition dipoles. Excitons over this network are not localised, but are shared between all the tryptophan molecules, in the same way that excitons are delocalised in the photosynthetic light-harvesting structures.

Photosynthesis absorbs light in the red and infra red. These forms of light are not available to tryptophan in proteins, but tryptophan is able to use ultra violet light emitted by the mitochondria. In fact, Tryptophan is sometimes referred to as chromophoric because of its ability to absorb UV light. Craddock implies that the same system that gives rise to quantum coherence in light-harvesting complexes could also give rise to it within the protein of neurons.

FUNCTIONAL QUANTUM STATES IN THE BRAIN

Following the recent papers discussed above, the debate on quantum coherence in living tissues has moved to a new stage. We now have definite evidence of functional quantum coherence in living matter, and also the existence of quantum entanglement, which may also possibly be functional. When this evidence is added to the similarities between the coherent structures in photosynthetic organisms and tryptophan, an amino acid that is common within neurons, we look to be moving into a zone where functional quantum states in the brain begin to look perfectly feasible.

QUANTUM AND CLASSICAL INTERACTION

The biologist, Stuart Kauffman, based at University of Vermont and & Tampere University, Finland (20.) is sceptical about ideas of consciousness based on classical and macroscopic physics. He proposes instead that consciousness is related to the border area between quantum and classical processing, where the non-algorithmic aspect of the quantum and the non-random aspect of the classical may be mixed. This is termed the 'poised realm', and is seen as applying to systems that include biomolecules and by extension brain systems.

THE POISED REALM

In rejecting the classical basis of mainstream consciousness studies, Kauffman instead proposes the idea of the 'poised realm', essentially the border of quantum and classical rules, which he suggests may support processing that is non-algorithmic, but at the same time non-random. This resembles the earlier non-algorithmic scheme proposed by Penrose. Kauffman puts forward the notion of a distinction between 'res potentia', the realm of the possible, or the quantum world, and 'res extensa' the realm of what actually exists, or the classical world. His proposal examines the meaning of the unmeasured or uncollapsed Schrödinger wave, and the question as to whether consciousness can participate at this level.

Kauffman discusses the modern quantum theory approach that distinguishes between an open quantum system and its environment. The open quantum system can be seen as the superposition of many possible quantum particles oscillating in phase. The information of the in-phase quanta can be lost through interaction with the environment, in the process known as decoherence. The information about the peaks and troughs of the Schrödinger wave, and the familiar interference pattern disappears, leading towards a classical system. The process of decoherence takes time, on a scale of one femtosecond. There is a problem regarding the physics of this, because while the mathematical description of the Schrödinger wave is time—reversible, decoherence has traditionally been treated as a time-irreversible dissipative process.

Recoherence: However, it is has in recent years become apparent that recoherence and the creation of a new coherence state is possible, with systems decohering to the point of being effectively classical, and then recohering. Classical information can itself produce recoherence. The Shor quantum error correction theorem shows that in a quantum computer with partially decoherent qubits, a measurement that injects information can bring the qubits back to coherence.

Kauffman, in collaboration with Gabor Vattay, a physicist at Eotvos University Budapest, and Samuli Niiranen, a computer scientist at

Tampere University worked out the concept of the 'poised realm' between quantum coherence and classical behaviour. It is in this poised region that Kaufmann suggests non-random, but also non-deterministic processes could arise. Between the open quantum system of the Schrödinger wave and classicality, there is an area that is neither algorithmic nor deterministic, and which is also acausal, and therefore unlike a classical computer. It is suggested that systems can hover between quantum and classical behaviour, this state being what Kaufmann refers to as the 'poised realm'. The non-deterministic processing in the 'poised realm' influences the otherwise deterministic processing of the classical sphere, which can in its turn alter the remaining quantum sphere. There is a two-way interaction between the quantum and classical region. The fact that this process deriving from the classical region is non-random introduces a non-random element into any remaining decoherence in the quantum system. Further, classical parts of the system can recohere, and inject classical information into the quantum system, thus introducing a degree of control into the superpositions of the quanta. In particular, the decision on which amplitudes reach the higher amplitudes, and thus have the greatest probability of decohering can be altered, thus altering the nature of particular classical outcomes.

This leads Kauffman on to discuss the recent discoveries in quantum biology, where quantum coherence and entanglement have been demonstrated in living photosynthetic organisms. The suggestion is that biomolecules are included in the systems that can hover between the quantum and the classical region, and further that this could apply not only to photosynthetic biomolecules, but also to biomolecules within neurons. Thus brain systems could be allowed to recohere to introduce further acausality into the system. Kaufmann views consciousness as a participation in res potentia and its possibilities. The presence of consciousness in the res potentia is also suggested to explain the lack of an apparent spatial location for consciousness. Qualia are suggested to be related to quantum measurement in which the possible becomes actual.

However, Kaufmann admits that all this still contains no real explanation of sensory experience. Kaufmann acknowledges that he is looking for something similar to Penrose, but thinks it may be located in the poised realm rather than in Penrose's objective reduction. Where the earlier scheme of Penrose still has the advantage is in the rounding off proposition that his objective reduction gives access to consciousness at the level of the fundamental spacetime geometry. Presumably Kaufmann assumes something of the kind. There is no particular reason why either quanta or classical structures or some mixture of them should be conscious, but we know that the quanta relate to fundamental properties such as charge and spin and to spacetime, and it seems reasonable on the same basis to look for consciousness as a fundamental property at this level.

SECTION 4

PENROSE & HAMEROFF

PENROSE—ALONE AS A DEEP THINKER ON CONSCIOUSNESS?

Roger Penrose is one of the very few thinkers to consider how consciousness could arise from first principles rather than merely trying to shoe horn it into nineteenth century physics. After having looked at the possible background to a fundamental theory of consciousness, in physics and biology, his ideas appear to be a good starting point from which to try to understand how consciousness might arise as a fundamental.

GÖDEL'S THEOREM

Penrose's approach was a counter attack on the functionalism of the late 20th century, which claimed that computers and robots could be conscious. He approached the question of consciousness from the direction of mathematics. The centre piece of his argument is a discussion of Gödel's theorem. The 20th century mathematician and logician, Kurt Gödel, demonstrated that any formal system or any significant system of axioms, such as elementary arithmetic, cannot be both consistent and complete. There will be statements that are undecidable, because although they are seen to be true, they are not provable in terms of the axioms.

PENROSE'S CONTROVERSIAL CLAIM

The Gödel theorem as such is not controversial in relation to modern logic and mathematics, but the argument that Penrose derived from it has proved to be highly contentious. Penrose claimed that the fact that human mathematicians can see the truth of a statement that is not demonstrated by the axioms means that the human mind contains some function that is not based on algorithms, and therefore could not be replicated by a computer. This is because the functioning of computers is based solely on algorithms (systems of calculations). Penrose therefore claimed that Gödel had demonstrated that human brains could do something that no computer was able to do.

ARGUMENTS AGAINST PENROSE'S POSITION

Some critics of Penrose have suggested that while mathematicians could go beyond the axioms, they were in fact using a knowable algorithm present in their brains. Penrose contests this, arguing that all possible algorithms are defeated by the Gödel problem. In respect to arguments as to whether computers could be programmed to deal with Gödel propositions, Penrose accepts that a computer could be instructed as to the non-stopping property of Turing's halting problem. Here, a proposition that goes beyond the original axioms of the system is put into a computation. However, this proposition is not part of the original formal system, but instead relies on the computer being fed with human insights, so as to break out of the difficulty. So the apparently non-algorithmic insights are required to supplement the functioning of the computer in this instance.

AN UNKNOWABLE ALGORITHM

Penrose further discusses the suggestion of an unknowable algorithm that enables mathematicians to perceive the truth of statements. He argues that there is no escape from the knowability of algorithms. An unknowable algorithm means an algorithm, whose specification could not be achieved. But any algorithm is in principle knowable, because it depends on the natural numbers, which are knowable. Further, it is possible to specify natural numbers that are larger than any number needed to specify the algorithmic action of an organism, such as a human or a human brain.

MATHEMATICAL ROBOTS

Penrose says that with a mathematical robot, it would not be practical to encode all the possible insights of mathematicians. The robot would have to learn certain truths by studying the environment, which in its turn is assumed to be based on algorithms. But to be a creative mathematician, the robot will need a concept of unassailable truth, that is a concept that some things are obviously true.

This involves the mathematical robot having to perceive that a formal system 'H' implies the truth of its Gödel proposition, and at the same time perceiving that the Gödel proposition cannot be proved by the formal system 'H'. It would perceive that the truth of the proposition follows from the soundness of the formal system, but the fact that the proposition cannot be proved by the axioms also derives from the formal system. This would involve a contradiction for the robot, since it would have to believe something outside the formal system that encapsulated its beliefs.

SOLOMON FEFERMAN

Amongst experts in this area who do not entirely reject Penrose's argument, Solomon Feferman has criticised Penrose's detailed argument (21.), but is much closer to his position than to that of mainstream consciousness studies. Feferman makes common cause with Penrose in opposing the computational model of the mind, and considering that human thought, and in particular mathematical thought, is not achieved by the mechanical application of algorithms, but rather by trial-and-error, insight and inspiration, in a process that machines will never share with humans. Feferman finds numerous flaws in Penrose's work, but at the end he informs his readers that Penrose's case would not be altered by putting right the logical flaws that Feferman has spent much time discovering.

Feferman's own position is that the computational-mind argument is misleading in terms of the weight that it places on the equivalence between Turing machines and formal systems. The model of mathematical thought in terms of formal systems is considered to be closer to the nature of human thought, and particularly mathematical thought, than to the functioning of Turing machines. The Turing machine model would assume that given a problem, human reason would plug away, applying the same algorithm indefinitely, in the hope of finding an answer.

Feferman says that it is ridiculous to think that mathematics is performed in this way. Trial-and-error reasoning, insight and inspiration, based on prior experience, but not on general rules, are seen as the basis of mathematical success. A more mechanical approach is only appropriate, after an initial proof has been arrived at. Then this approach can be used for mechanical checking of something initially arrived at by trial-and-error and insight. He views mathematical thought as being non-mechanical. He says that he agrees with Penrose that understanding is essential to mathematical thought, and that it is just this area of mathematical thought that machines cannot share with us.

PENROSE'S SEARCH FOR A NON-ALGORITHMIC FEATURE

Penrose went on to ask, what it was in the human brain that was not based on algorithms. The physical law is described by mathematics, so it is not easy to come up with a process that is not governed by algorithms. The only plausible candidate that Penrose could find was the collapse of the quantum wave function, where the choice of the position of a particle is random, and therefore not the product of an algorithm. However, he considered that the very randomness of the wave collapse disqualifies it as a useful basis for the mathematical judgement or understanding in which he was initially interested. This problem was to lead Penrose a new proposition for both physics and consciousness.

THE WAVE FUNCTION

In respect of consciousness, it is Penrose's attitude to the reality of the quantum wave function collapse that is the important area. In particular, he disagrees with the traditional Copenhagen interpretation, which says that the theory is just an abstract calculational procedure, and that the quanta only achieve objective reality when a measurement has been made. Thus in the Copenhagen approach reality somehow arises from the unreal or from abstraction, giving a dualist quality to the theory.

The discussion of quantum theory repeatedly comes back to the theme that Penrose regards the quantum world and the uncollapsed wave function as having objective existence. In Penrose's view, the objective reality of the quantum world allows it to play a role in consciousness. Penrose emphasises that the evolution of the wave function portrayed by the Schrödinger equation is both deterministic and linear. This aspect of quantum theory is not random. Randomness only emerges when the wave function collapses, and gives the choice of a particular position or other properties for a particle.

Penrose discusses the various takes made on wave function collapse by physicists. Some would like everything to depend on the Schrödinger equation, but Penrose rejects this idea, because it is impossible to see how the mechanism of this equation could produce the transformation from the superposition of alternatives, as found in the quantum wave, to the random choice of a single alternative.

He also discusses the suggestion that the probabilities of the quantum wave that emerges into macroscopic existence could arise from uncertainties in the initial conditions and that this system is analogous to chaos in macroscopic physics. This does not satisfy Penrose, who points out that chaos is based on non-linear developments, whereas the Schrödinger equation is linear.

IMPORTANT DISTINCTION BETWEEN PENROSE AND WIGNER

Penrose also disagrees with Eugene Wigner's suggestion that it is consciousness that collapses the wave function, on the basis that consciousness is only manifest in special corners of spacetime. Penrose himself advances the exact opposite proposal that the collapse of a special (objective) type of wave function produces consciousness. It is important to stress this difference between the Penrose and the Wigner position, as some commentators mix up Wigner's idea with Penrose's propositions on quantum consciousness, and then advance a refutation of Wigner, mistakenly believing it to be a refutation of Penrose.

Penrose is also dismissive of the 'many worlds' version of quantum theory, which would have an endless splitting into different universes with, for instance, Schrödinger's cat alive in one universe and dead in another universe. Penrose objects to the lack of economy and the multitude of problems that might arise from attempting such a solution, and in addition argues that the theory does not explain why the splitting has to take place, and why it is not possible to be conscious of superpositions.

OBJECTIVE REDUCTION

Penrose instead argues for some new physics, and in particular an additional form of wave function collapse. If the superpositions described by the quantum wave extended into the macroscopic world, we would in fact see superpositions of large-scale objects. As this does not happen, it is argued that something that is part of objective reality must take place to produce the reality that we actually see. This requirement for new physics is often criticised as unjustified. However, these criticisms tend to ignore the fact that while quantum theory provides many accurate predictions, there has never been satisfactory agreement about its interpretation, nor has its conflict with relativity been resolved.

CONSCIOUSNESS, SPACETIME, THE SECOND LAW & GRAVITY

Penrose sees consciousness as not only related to the quantum level but also to spacetime. He discusses the spacetime curvature described in general relativity. He looks at the effect of singularities relative to two spacetime curvature tensors, Weyl and Ricci. Weyl represents the tidal effect of gravity, by which the part of a body nearest to the gravitational source falls fastest creating a tidal distortion in the body. Ricci represents the inward pull on a sphere surrounding the gravitational force. In a black hole singularity, the tidal distortion of Weyl would predominate over Ricci, and Weyl goes to infinity at the singularity.

However, in the early universe expanding from the Big Bang, the tidal distortion is absent, so Weyl=0, while it is the inward pressure of Ricci that predominates. So the early universe is seen to have had low entropy with Weyl close to zero. Weyl is related to gravitational distortions, and Weyl close to zero indicates a lack of gravitational clumping, just as Weyl at infinity indicated the gravitational collapse into a black hole. Weyl close to zero and low gravitational clumping therefore indicate low entropy at the beginning of the universe.

The fact the Weyl is constrained to zero is seen by Penrose as a function of quantum gravity. The whole theory is referred to as the Weyl curvature hypothesis. The question that Penrose now asks is as to why initial spacetime singularities have this structure. He thinks that quantum theory has to help with the problem of the infinity of singularities. This would be a quantum theory of the structure of spacetime, or in other words a theory of quantum gravity.

Penrose regards the problems of quantum theory in respect of the disjuncture between the Schrödinger equations deterministic evolution and the randomness in wave function collapse as fundamental. He thinks in terms of a time-asymmetrical quantum gravity, because the universe is time-asymmetric from low to high entropy. He argues that the conventional process of collapse of the wave function is time-asymmetric. He describes an experiment where light is emitted from a source and strikes a half-silvered mirror with a resulting 50% probability that the light reaches a detector and 50% that it hits a darkened wall. This experiment cannot be time reversed, because if the original emitter now detects an incoming photon, there is not a 50% probability that it was emitted by the wall, but instead 100% probability that it was emitted by the other detecting/emitting device.

Penrose relates the loss of information that occurs in black holes to the quantum mechanical effects of the black hole radiation described by Stephen Hawking. This relates the Weyl curvature that is seen to apply in black holes and the quantum wave collapse. As Weyl curvature is related to the second law of thermodynamics, this is taken to show that the quantum wave reduction is related to the second law and to gravity.

Penrose proposes that in certain circumstances there could be an alternative form of wave function collapse. He called this objective reduction (OR). He suggests that as a result of the evolution of the Schrodinger wave, the superpositions of the quanta grow further apart. According to Penrose's interpretation of general relativity, each superposition of the quanta is conceived to have its own spacetime geometry. The separation of the superpositions, each with its own spacetime geometry constitutes a form of blister in space-time. However once the blister or separation grows to more than the Planck length of 10^{-35} metres, the separations begin to be affected by the gravitational force, the superposition becomes unstable, and it soon collapses under the pressure of its gravitational self-energy. As it does so, it chooses one of the possible spacetime geometries for the particle. This form of wave function collapse is proposed to exist in addition to the more conventional forms of collapse.

EVIDENCE FOR NON-COMPUTATIONAL SPACETIME

In support of this, he points out that when the physicists, Geroch and Hartle, studied quantum gravity, they ran up against a problem in deciding whether two spacetimes were the same. The problem was solvable in two dimensions, but intractable in the four dimensions that accord with the four dimensional spacetime, in which the superposition of quantum particles needs to be modelled. It has been shown that there is no algorithm for solving this problem in four dimensions.

Earlier the mathematician, A. Markov, had shown there was no algorithm for such a problem, and that if such an algorithm did exist, it could solve the Turing halting, for which it had already been shown that there was no algorithm. The possibly non-computable nature of the structure of four-dimensional space-time is deemed to open up the possibility that wave function collapses could give access to this non-computable feature of fundamental space-time.

TESTING PENROSE'S OBJECTIVE REDUCTION

A long-term experiment is underway to test Penrose's hypothesis of objective reduction. This experiment is being run by Dirk Bouwmeester at the University of California, Santa Barbara and involves mirrors only ten micrometres across and weighing only a few trillionths of a kilo, and the measurement of their deflection by a photon. The experiment is expected to take ten years to complete. This means that theories of consciousness based on objective reduction are likely to remain speculative for at least that length of time. However, the ability to run an experiment that could falsify objective reduction, at least qualifies it as a scientific theory.

SIGNIFICANCE FOR CONSCIOUSNESS

The significance of this for the study of consciousness is that, in contrast to the conventional idea of wave function collapse, this form of collapse is suggested to be non-random, and instead driven by a non-computable function at the most fundamental level of spacetime. Penrose argues that, in contrast to the conventional wave function form of collapse, there are indications that in this case, there is a decision process that is neither random nor computationally/ algorithmically based, but is more akin to the 'understanding' by which Penrose claims the human brain goes beyond what can be achieved by a computer.

HAMEROFF

When Penrose first proposed his ideas on consciousness, he had no significant suggestion as to how this could be physically instantiated in the brain. Subsequent to this, Stuart Hameroff proposed a scheme by which he thought Penrose's concept of objective reduction might be instantiated in neurons, giving rise to the theory of orchestrated objective reduction (Orch OR).

SINGLE-CELLS ORGANISMS AND NEURONS

Hameroff emphasises that single-cell organisms have no nervous system, but can perform complicated tasks, which could only be achieved by means of some form of internal processing. He surmised that the same form of processing could exist in brain cells. Thus Hameroff viewed each neuron as a computer. Within the neuron, a number of areas such as the ion channels and parts of the synapses were considered as possible sites for information processing and ultimately consciousness. However, another candidate, the cytoskeleton, came to be viewed as the component of the neuron best suited to information processing. The cytoskeleton comprises a protein scaffolding that provides a structural support for all living cells including neurons.

MICROTUBULES

Microtubules are the major element of the cytoskeleton. As well as providing structural support for the cell, they are important for internal transport, including the transport of neurotransmitter vesicles to synapses in neurons. Hameroff suggested that microtubules were suitable for information processing, and in addition to this that they could support quantum coherence and the objective reduction looked for in Penrose's theory. The microtubules are comprised of the protein tubulin, which is made up of an alpha and beta tubulin dimer. The microtubules are formed of 13 filamentous tubulin chains skewed so that the filaments run down the cylinder of the microtubule in a helical form, and hexagonal in that each tubulin dimer has six neighbours. Each turn of this helix is formed by thirteen dimers, and creates a slightly skewed hexagonal lattice, considered to be suitable for information processing. The intersections of the windings of the protofilaments are also the attachment sites for microtubular associated proteins (MAPs) that help to bind the cytoskeleton together.

The nature and activity of microtubules in neurons is markedly different from that in other body cells. Neuron microtubules are denser and more stable than those in other cells. In neurons microtubules are also more important for linking parts of the cell, such as taking synaptic vesicles from the Golgi apparatus in the cell body down to the axon terminal, and carrying protein and RNA to the dendritic spines.

SUITABILITY FOR INFORMATION PROCESSING

It is the geometry of this lattice based on tubulin sub-units that is considered to have a potential for information processing. Within the cylindrical lattice of the microtubule, each tubulin is in a hexagonal relationship, by virtue of being in contact with six other neighbouring tubulins. Each dimer would be influenced by the polarisation of six of its neighbours, which in turn makes them suitable for the transmission of signals. It is suggested that tubulins could interact with neighbouring tubulin by means of dipole interactions. The dipole-coupled conformation for each tubulin could be determined by the six surrounding tubulins.

The geometry of a quantum computing lattice could be suitable for quantum error correction. This latter view is consistent with the recent studies of photosynthetic systems. The intersections of the windings of the protofilaments are also the attachment sites for microtubular associated proteins (MAPs) that help to bind the cytoskeleton together.

Hameroff describes protein conformation as a delicate balance between contervailing forces. Proteins are chains of amino-acids that fold into three dimensional conformations. Folding is driven by van der Waals forces between hydrophobic amino-acid groups. These groups can form hydrophobic pockets in some proteins. These pockets are critical to the folding and regulation of protein. Amino acid side groups in these pockets interact by van der Waals forces.

DENDRITES AND CONSCIOUSNESS

Hameroff related consciousness not to the axons of neurons that allow forward communication with other neurons, but the dendrites that receive inputs from other neurons. The cytoskeleton of the dendrites is distinct both from that found in cells outside the brain, and also from the cytoskeleton found in the axons of neurons. The microtubules in dendrites are shorter than those in axons and have mixed as opposed uniform polarity. This appears a sub-optimal arrangement from a structural point of view, and it is suggested that in conjunction with microtubule associated proteins (MAPs), this arrangement may be optimal for information processing.

These microtubule/MAP arrangements are connected to synaptic receptors on the dendrite membrane by a variety of calcium and sodium influxes, actin and other inputs. Alterations in the microtubule/MAPs network in the dendrites correlate with the rearrangement of dendritic synapatic receptors. Hameroff points out that changes in dendrites can lead to increased synaptic activity. The changes in dendrites involve the number and arrangement of receptors and the arrangement of dendritic spines and dendrite-to-dendrite connections. The main function of dendrites is seen to be the handling of signal input into the neuron, which may eventually result in an axon spike.

DENDRITIC SPINES, THE DENDRITIC CYTOSKELETON & INFORMATION TRANSMISSION

Neurons receive inputs through dendrites and dispatch signals through axons. Dendritic spines are the points at which signals from other neurons enter the dendrites. There is evidence for interactivity between dendritic spines and the dendritic cytoskeleton. The connection between the membrane and the cytoskeleton has tended to be ignored. Actin filaments are concentrated in dendritic spines and near to axon terminals. These bind to scaffolding proteins and interact with signalling molecules. There are also interactions between ion channels and the cytoskeleton, especially actin filaments. Experimental work suggests that the cytoskeleton and actin filaments in particular can regulate ion channels that are part of basic neural processing. Recent studies indicate cross-linker proteins between actin filaments and microtubules, in additions to MAP 2 and tau which are known to bind to actin filaments. The dendritic spines can be modulated by actin indicating that cytoskeletal proteins can influence synaptic plasticity. The spines receive glutamate inputs by means of NMDA and AMPA receptors. Actin holds signal transduction molecules close to the NMDA receptors, and this links these receptors to signal cascades within the neuron. Actin is also important for anchoring ion channels, and congregating them in clusters. Actin filaments are known to control the excitability of some ion channels, such as the K+ channel, and it also binds to the Na^+ and Ca^{2+} ion channels.

Scaffolding proteins such as the post-synaptic density protein, PSD95 and gephyrin a GABA receptor scaffolding protein, secure the membrane receptors in the dendritic spine, and attach them to protein kinases and also to actin filaments that constitute part of the cytoskeleton. Gephyrin concentrates GABA receptors at post synaptic sites, while actin filaments support the movements of geyphrin complexes. Actin filaments are concentrated immediately below the neuronal membrane, but also penetrate into the rest of the cytoskeleton and are heavily concentrated in dendritic spines. The actin filaments are shown to be involved in the reorganisation of dendritic spines following stimulation. They also hold in place receptors, ion channels and transduction molecules.

CONDITIONS FOR COHERENT VIBRATIONS IN THE CYTOSKELETON

J. Pokorny of the Academy of Sciences of the Czech Republic published a paper on coherent activity in the cytoskeleton in Bioelectrochemistry and Bioenergetics (1999)

Prokorny suggests that the polarity of vibrational structures in organic matter can lead to coherent states and energy condensation. Microtubules are suggested to satisfy this requirement. He stated that that mechanisms governing the high degree of organisation in living organisms are largely unknown. The type of spatial order seen with inorganic crystalline substances is not apparent in organic matter, such as microtubules and actin filaments, and it is generally agreed that the self assembly and ordering of living matter is based on mechanisms peculiar to biological systems.

Biological matter has polar properties suggesting that there are electromagnetic aspects to its organisation.

Prokorny discusses at some length proposals by Frohlich, who postulated that long-range quantum mechanical phase correlations could exist in biological systems. His ideas were based on three concepts, the polar nature of biological structures, energy supply to the system, and energy transfer between oscillators within the system. Some modes of motion could become strongly excited, and remain far from thermal equilibrium. The polar nature of biological objects made it likely that this would lead to longitudinal oscillations. His calculations suggested that these processes could produce a form of quantum coherence capable of supporting dissipationless transfer of energy.

Pokorny argues that microtubules have the right structure to generate this type of energy. The tubulin sub-components are electrical dipoles, and the microtubules as a whole are polar structures, with positive and negative ends. Vibrations in such a polar system give rise to an electromagnetic field. Energy is exchanged between the vibrating structure and the region around it.

It is suggested that some of the energy involved in conformational changes in the tubulin proteins of microtubules could be transformed into vibrational energy. The altered forces between the protein subunits of the microtubule would supply energy to the whole microtubule lattice. Pokorny considers it possible that polarisation waves may become coherent on the basis of Frohlich's calculations. It is thought possible that polar vibrations do work in the vicinity of the microtubule. Pokorny further suggests that the electromagnetic activity observed in living cells such as yeast cells may be relevant to the activities in microtubules.

Prokorny produced a further paper in 2004 titled 'Excitation of vibrations in microtubules in living cells'. This paper argues that an ion layer and bound water should allow energy created inside microtubules to drive excitations. The paper starts by stating that the polar (opposite electric charge) character of biological objects suggests that there are longitudinal oscillations. The author's calculation show that some forms of such energy are not thermalised, but are instead condensed into a pattern of oscillations. In this context, the structure of the cell's cytoskeleton, which is based on microtubules, is stated to satisfy the basic requirements for an oscillating electric field (1. Pokorny, J, 1999). Later work by the author (2. Pokorny, J, 2003) confirmed the possibility of oscillations despite the existence of surrounding water.

Some of Prokorny's concepts were experimentally tested (3. Pokorny, J. et al, 2001). In this experiment recorded electromagnetic emissions from cells were regarded as promising, although more accurate tests would be needed to show whether the emissions came from microtubules rather than other parts of the cell.

HAMEROFF AND π ELECTRON CLOUDS

Where Hameroff moves on to discussing π electron clouds, he comes closer to the type of functional quantum coherence identified in photosynthetic systems. There has been much discussion over the last two decades as to how microtubules or any other structure in the neurons could sustain quantum states for long enough for them to be relevant to neural processing. In the light of more recent studies of quantum coherence in photosynthetic systems, it looks most likely that any quantum coherence in microtubules would relate to π electron clouds.

Mainstream research moved away from the idea of quantum processes in living organisms during the second half of the 20th century, although a few physicists such as Fröhlich kept the idea alive. Fröhlich proposed that biochemical energy could pump quantum coherent dipole states in geometrical arrays of non-polar π electron delocalised clouds. Such electron clouds are now known to be isolated from water and ions, and present in cells within membranes, microtubules and organelles. These electron clouds can use London forces, involving interaction between instantly forming dipoles in different electron clouds, to govern the conformation of biomolecules, including proteins.

AROMATIC RINGS

Life is based on carbon chemistry and notably carbon ring molecules, such as benzene, which has delocalised electron clouds in which London forces are active. Carbon has four atoms in its outer shell, able to form four covalent bonds with other atoms. In some cases two of the electrons form a double bond with another atom, and the remaining two outer electrons remain mobile and are known as π electrons. In benzene, there are three double bonds between six carbon atoms, such that all six carbon atoms are involved in a bond. The ring structure, into which these atoms are formed, famously came to its discoverer, Friedrich von Kekule, in a dream of a snake biting its tail. The π electrons are delocalised in this system. Benzene rings and the more complex indole rings are referred to as aromatic rings, and make up several of the amino acid side groups that are attached to proteins.

PROTEIN FOLDING AND π ELECTRON CLOUDS

Proteins constitute the driving machinery of living systems, since it is they which open and close ion channels, grasp molecules to enzymes and receptors, make alterations within cells, and govern the bending and sliding of muscle filaments. The organisation of protein is still poorly understood. Proteins are formed from 20 different amino acids with an enormous number of possible sequences. Van der Waals forces are involved in the proteins folding into different conformations, with a huge number of possible patterns of attraction and repulsion between the side groups of the protein.

During the protein folding process there are non-local interactions between aromatic rings, which has been seen as suggestive of quantum mechanical sampling of possible foldings. Once formed a protein structure can be stabilised by outwardly facing polar groups and by regulation from non-polar regions within. The coalescence of non-polar amino acid side groups, such as two aromatic rings, can result in extended electron clouds constituting hydrophobic pockets. Protein conformation represents a delicate balance between forces such as chemical and ionic bonds, and as a result London forces driven by π electrons in hydrophobic pockets can tip the balance and thus govern conformations of protein.

HYDROPHOBIC POCKETS AND ENTANGLEMENT

The more solid parts of cells include protein structures, and these have within them hydrophobic areas containing hydrophobic or oil-like molecules with delocalised π electron clouds. In water, non-polar oily molecules such as benzene, which are hydrophobic are pushed together, attracting each other by London forces, and eventually aggregate into stable regions shielded from interaction with water. London forces can govern the configurations of protein in these regions. Such regions occur as pockets in proteins. In the repetitive structures of the tubulin dimer, π electrons clouds may be separated by less than two nanometres, and this is seen as conducive to entanglement, electron tunnelling or exciton hopping between dimers and connections between the electron clouds extending down the length of the neuron.

Tubulin has a dimer form with an alpha and beta monomer joined by a 'hinge'. The tubulin has a large non-polar region in the beta monomer just below the 'hinge'. Other smaller non-polar regions with π electron rich indole rings, are distributed throughout the tubulin with distances of about two nanometres between them. The positioning of π electron clouds within about two nanometres of one another is suggested to allow the electrons to become entangled. This entanglement could spread through the microtubule and to other microtubules in the same dendrite.

Following on recent research, it has become possible to compare the situation in microtubules to quantum coherence and entanglement in photosynthetic organisms, something unknown when researchers such as Tegmark argued against the possibility of functionally relevant quantum coherence in the brain. In photosynthetic systems light-harvesting chromophore molecules use dipoles to provide 99% efficiency in energy transfer from the light harvesting antennae to the reaction centre. The studies show that instead of quantum coherence being destroyed by the environment within the organism, a limited amount of noise in the environment acts to drive the system.

TRYPTOPHAN

Similar models can be used for chromophores in photosynthetic systems and for tryptophan, an aromatic amino acid that is one of the 20 standard amino acids making up protein. Tryptophan has eight molecules extending over the length of the tubulin protein dimer, and it possesses strong transition dipoles. Excitons over this network are not localised, but are shared between all the tryptophan molecules, in the same way that excitons are delocalised in the photosynthetic light-harvesting structures.

Photosynthesis absorbs light in the red and infra red. These forms of light are not available to tryptophan in proteins, but tryptophan is able to use ultra violet light emitted by the mitochondria. In fact Tryptophan is sometimes referred to as chromophoric because of its ability to absorb UV light. It is becoming feasible to suggest that the same system that gives rise to quantum coherence in light-harvesting complexes could also give rise to it within the protein of neurons.

PENROSE & HAMEROFF (2011)

In their latest joint paper published as a chapter in Consciousness and the Universe (2011) (22.) Penrose and Hameroff deal with aromatic rings and proposed hydrophobic channels within microtubules that could be crucial for a quantum theory of consciousness. They point to unexpected discoveries in biology. The most important change since Penrose and Hameroff first propounded their ideas in the 1980s and 1990s is the recent discoveries in biology relative to higher temperature quantum activity. In 2003 Ouyang & Awschalom showed that quantum spin transfer in phenyl rings (an aromatic ring molecule like those found in protein hydrophobic pockets) increases at higher temperatures. In 2005 Bernroider and Roy (23.) researched the possibility of quantum coherence in K+ neuronal ion channels.

A more crucial discovery came in 2007 when it was demonstrated that quantum coherence was functional in efficiently transferring energy within photosynthetic organisms (Engel et al, 2007). Subsequent papers showed functional quantum coherence in multicellular plants and also at room temperature. In 2011 papers by Gauger et al and Luo and Lu dealt with higher temperature coherence in bird brain navigation and in protein folding. Work by Anirban Bandyopadhyay with single animal microtubules showed eight resonance peaks correlated with helical pathways round the cylindrical microtubule lattice. This allowed 'lossless' electrical conductance.

TUBULIN, AND AROMATIC RINGS: BUILDING BLOCKS OF CONSCIOUSNESS?

Each tubulin protein contains the amino acids tryptophan and phenylalanine with aromatic rings. Each hydrophobic pocket in the tubulin is suggested to be composed of four such aromatic rings, with the hydrophobic pockets being arranged in channels. Van der Waals London forces operate in the hydrophobic pockets in tubulin, based on the π electron rings of tryptophan and phenylaline. This concept derives originally from Fröhlich, who suggested that proteins are synchronised by the oscillations of dipoles in the electron clouds of these amino acids. Anaesthetic gases are similarly suggested to work through their action on aromatic amino acids in hydrophobic pockets in neuronal proteins, including membrane proteins.

HYDROPHOBIC CHANNELS AND LONG-RANGE VAN DER WAALS FORCES

A paper published in 1998 (Nogales et al) described the structure of the tubulin protein and identified the existence and location of the non-polar aromatic amino acids tryptophan and phenylamine in tubulin. These are located in hydrophobic pockets, but these pockets are within 2 nanometres of one another, and collectively they can be interpreted as hydrophobic channels or pathways rather than mere pockets. This is suggested to allow linear arrays of electron clouds capable of supporting long-range van der Waals London forces. The quantum channels in individual tubulins are seen as being aligned with those in neighbouring tubulins within the microtubule lattice, and these provide helical winding patterns.

The authors also make a direct reply to one critic in particular (McKemmish et al, 2010) McKemmish claimed that switching between two states of the tubulin protein in the microtubules would involve conformational changes requiring GTP hydrolysis which in turn would involve an impossible energy requirement. The authors however claim that electron cloud dipoles (van der Waals London forces) are sufficient to achieve switching without large conformational changes.

CRITICISMS OF THE HAMEROFF SCHEME

Where the Hameroff version of quantum consciousness remains ambitious relative to existing scientific knowledge is in the proposed link to the global gamma synchrony, the brain's most obvious correlate of consciousness. Hameroff proposes that coherence within dendrites connects via gap junctions to other neurons and thus to the neuronal assemblies involved in the global gamma synchrony. He thus proposes the existence of quantum coherence over large areas of the brain, sometimes including multiple cortical areas and both hemispheres of the brain.

Hameroff pointed to gap junctions as an alternative to synapses for connections between neurons. Neurons that are connected by gap junctions depolarise synchronously. Cortical inhibitory neurons are heavily studded with gap junctions, possibly connecting each cell to 20 to 50 other. The axons of these neurons form inhibitory GABA chemical synapses on the dendrites of other interneurons. Studies show that gap junctions mediate the gamma synchrony. On this basis, Hameroff suggested that cells connected by gap junctions may in fact constitute a cell assembly, with the added advantage of synchronous excitation. In this scheme computations are suggested to persist for 25 ms, thus linking them to the 40Hz gamma synchrony.

The attempt to extend a proposal for quantum features from single neurons out to neuronal assemblies of millions of neurons resurrects the nay-sayer objections concerning time to decoherence. The photosynthetic states that have been demonstrated persist for only femtosecond and picosecond timescales. Where the decoherence argument still stands up is in dealing with a system that needs to be sustained for 25 milliseconds. Further to this the Hameroff's gamma wide theory involves difficult arguments about the ability of coherence to pass from neuron to neuron via the gap junctions.

Danko Georgiev, a researcher at Kanazawa University also criticises Hameroff's requirement for microtubules to be quantum coherent for 25 ms. This has been generally regarded as an ambitious timescale for quantum coherence, and Georgiev objects on the grounds that enzymatric functions in proteins take place on a very much quicker 10-15 picosecond timescale. Georgiev wants to base his version of OR consciousness on this 10-15 picosecond timescale. Such a rapid form of objective reduction would also remove the necessity for the gel-sol cycle to screen microtubules from decoherence, as it does in the Hameroff version of objective reduction.

AXONS, DENDRITES AND SYNAPSES

Georgiev also criticises Hameroff's emphasis on conscious processing as being concentrated in the dendrites. He claims that Hameroff's does not allow any consciousness in axons, and this creates a problem in explaining the problematic firing of synapses. Only 15-30% of axon spikes result in a synapse firing, and it is not clear what determines whether or not a synapse fires. He discusses the probabilistic nature of neurotransmitter release at the synapses, and the possible connection this has with quantum activity in the brain.

Georgiev points out that an axon forms synapses with hundreds of other neurons, and that if the firing of all these synapses was random, the operation of the brain could prove lacking in organisation.

He suggests instead the choice of which synapses will fire is connected to consciousness, and that consciousness acts within neurons. Each synapse has about 40 vesicles holding neurotransmitters, but only one vesicle fires at any one time. Again the choice of vesicle seems to require some form of ordering. The structure of the grid in which the vesicles are held is claimed to be suitable to support vibrationally assisted quantum tunnelling.

Georgiev's emphasises the onward influence of solitons (quanta propagating as solitary waves) from the microtubules to the presynaptic scaffold protein, from where, via quantum tunnelling, they are suggested to influence whether or not synapses fire in response to axon spikes. Jack et al (1981) suggested an activation barrier, restricting the docking of vesicles and the release of neurotransmitters. The control of presynaptic proteins is suggested to overcome this barrier, and to regulate the vesicles that hold neurotransmitters in the axon terminals. This is suggested to be the process that decides whether a synapse will fire in response to an axon spike, and if it does, which of a choice of 40 or so vesicles will release its neurotransmitters.

The system he describes involves the neuronal cytoskeleton, and particularly the pre and post-synaptic scaffold proteins. Here, it is suggested that consciousness arises from the objective reduction of the wave function within these structures. The timescale of the system is argued to be defined by changes in tubulin conformations within the cytoskeleton and by the enzyme action in the scaffold proteins, which involves a timescale of 10-15 picoseconds, and thus implies a decoherence time on the same scale. Georgiev points out that it is much easier to suppose a decoherence time of this length in the brain than the 25 ms demanded by the Hameroff proposals.

ION CHANNELS AND CONSCIOUSNESS—GUSTAV BERNROIDER

As an aside from microtubules Gustav Bernroider at Salzburg University has proposed a quantum information system in the brain that is driven by the entangled ion states in the voltage-gated ion channels of the membranes of neurons. These ion channels, situated in the neuron's membrane are a crucial component of the conventional neuroscience description of axon spiking leading to neural transmitter release at the synapses. The ion channels allow the influx and outflux of ions from the cell driving the fluctuation of electrical potential along the axon, which in turn provides the necessary signal to the synapse.

The work concentrates attention on the potassium (K+) channel and in particular the configuration of this channel when it is in the closed state. This channel is traditionally seen as having the function of resetting the membrane potential from a firing to a resting state. This is achieved by positively charged potassium (K+) ions flowing out of the neuron through the channel.

Progress in atomic-level spectroscopy of the membrane proteins that constitute the ion channels and the accompanying molecular dynamic simulations indicate that the organisation of the membrane proteins carries a logical coding potency, and also implies quantum entanglement within ion channels and possibly also between different ion channels. An increasing number of studies show that proteins surrounding membrane lipids are associated with the probabilistic nature of the gating of the ion channels (Doyle, 1998, Zhou, 2001, Kuyucak, 2001).

This theory draws particularly on the work of MacKinnon and his group, notably his crystallographic X-ray work. The study shows that ions are coordinated by carboxyl based oxygen atoms or by water molecules. An ion channel can be in either a closed or an open state, and in the closed state there are two ions in the permeation path that are confined there. This closed gate arrangement is regarded as the essential feature with regard to their research work. The open gate presents very little resistance to the flow of potassium ions, but the closed gate is a stable ion-protein configuration.

The ion channel serves two functions, selecting K+ ions as the ones that will be given access through the membrane, and then voltage-gating the flow of the permitted K+ ions. In the authors' view, recent studies also require a change in views both of the ion permeation and of the voltage-gating process. A charge transfer carried by amino acids is involved in the gating process. In the traditional model the charges were completely independent, whereas in the new model there is coupling with the lipids that lie next to the channel proteins. This view, which came originally from MacKinnon, is now supported by later recent studies. The authors think that the new gating models are more likely to support computational activity, than were the traditional models.

Three potassium ions are involved in the ion channel's closed configuration. Two of these are trapped in the permeation path of the protein, when the channel gate is closed. The filter region of the ion channel is indicated by the recent studies to have five binding pockets in the form of five sets of four carboxyl related oxygen atoms. Each of the two trapped potassium ion are bound to eight of the oxygen atoms, i.e. each of them are bound to two out of the five binding pockets. The author's calculations predict that the trapped ions will oscillate many times before the channel re-opens, and the calculations also suggest an entangled state between the potassium ions and the binding oxygen atoms. This structure is seen as being delicately balanced and sensitive to small fluctuations in the external field. This sensitivity is viewed as possibly being able to account for the observed variations in cortical responses.

The theory also relates the results of recent studies of the potassium channel and its electrical properties to the requirements for quantum computing. There have been schemes for quantum computers involving ion traps, based on electostatic interactions between ions held in microscopic traps, that have a resemblance to Bernroider's interpretation of the possible quantum state of the K+ channel.

The authors deny that the rapid decoherence of quantum states in the brain calculated by Tegmark applies to their model. They argue that the ions are not freely moving in the ion filter area of the closed potassium channel, but are held in place by the surrounding electrical charges and the external field. The ions are particularly insulated within the carboxyl binding pockets, and it is suggested the decoherence could be avoided for the whole of the gating period of the channel, which is in the range of 10-13 seconds.

The authors also raise the question of whether given quantum coherence in the ion channel, it is possible for the channel states to be communicated to the rest of the cell membrane. This could include connections to other ion channels in the same membrane, possibly by means of quantum entanglement.

Bernroider's work might not be considered to be a fully fledged separate quantum consciousness theory. In the early part of the decade, Bernroider seemed to associate himself with David Bohm's implicate order, but the lack of much specific neuroscience in Bohm's version makes it hard to make any definite connection between it and the type of detailed neuroscientific argument offered by Bernroider. In the light of the advances in biology, and the potential for coherence to be supported by aromatic molecules within microtubules, it might be feasible to suggest that quantum coherence in the ion channels works together with coherence in other parts of the neuron.

SECTION 5

BRAIN PROCESSING AND CONSCIOUSNESS

We have looked at the question of consciousness as a fundamental in terms of the quanta and spacetime, and we have looked at the possibility of quantum states in the brain. This brings us to the further question of how consciousness is related to larger-scale brain processing.

CONSCIOUS & NON-CONSCIOUS PROCESSING

Twentieth century consciousness studies tended to be very insistent that consciousness was the product of neural systems as described in text books that made no distinction between conscious and non-conscious processing. Any system that did what neurons did would produce consciousness, or alternatively consciousness was what it was like to have a brain, without any distinction between different parts of the brain. More recent research is indicative of consciousness arising in specific areas of the brain, and that on a transient basis. While this cannot be said to disprove the twentieth century claims, it does at least suggest a more careful and discriminating approach to understanding what gives rise to consciousness.

Researchers, Goodale and Milner (24.) points to a ventral stream in the brain that produces conscious visual perception, and to a separate dorsal stream supporting non-conscious visuomotor orientations and movements. Goodale and Milner refer to a patient, who had suffered brain damage as a result of an accident. She had difficulty in separating an object from its background, which is a crucial step in the process of visual perception. She could manipulate images in her mind, but could not perceive them directly on the basis of signals from the external world, thus demonstrating that images generated by thought use a different process from direct perceptions of the external world. In modern neuroscience, perception is not viewed as a purely bottom-up process resulting from analysis of patterns of light, but is seen as also requiring a top-down analysis based on what we already know about the world.

The patient's visual problems seemed paradoxical. If a pencil was held out in front of her, she couldn't tell what it was, but she could reach out and position her hand correctly to grasp it. This contrast between what she could perceive and what she could do was apparent in a number of other instances.

The researchers saw the patient's state as indicative of the existence of two partly independent visual systems in the brain, one producing conscious perception, and the other producing unconscious control of actions. They point to instances of patients with the opposite of their patient's problems, who are able to perceive objects, their size and location, but are unable to translate this into effective action. These patients may be able to accurately estimate the size of an object, but are unable to scale their grip in taking hold of it, despite having no underlying problem in their movement ability. These patients with exactly the opposite problems from the first patient are taken to suggest partly-independent brain systems, one supporting perception, and the other supporting vision based action.

It has been found that even in more primitive organisms, there can be separate systems for catching prey, and for negotiating obstacles, with distinct input and output paths. This modularity is also found in the visual systems of mammals. The retina projects to a number of different regions in the brain. In humans and other mammals, the two most important pathways are to the lateral geniculate nucleus in the thalamus and the superior collicus in the midbrain. The path to the superior collicus is the more ancient in evolutionary terms, and is present in more primitive organisms. The pathway to the geniculate nucleus is more prominent in humans. The geniculate nucleus projects in turn to the primary visual cortex or V1. The mechanisms that generate conscious visual representations are recent in evolutionary terms, and are seen as distinct from the visuomotor systems, which are the only system available to more primitive organisms.

The perceptual system is not seen as being specifically linked to motor outputs. The perceptual representation may be additionally shaped by emotions and memories, as well as the immediate light signals from the environment. By contrast, visuomotor activity may be largely bottom up, drawing on the analysis of light signals, and is not accessible to conscious report. As such, this appears little different from the systems used by primitive organisms, while perception is a product of later evolution. Perceptual representations of the external world have meaning, and can be used for planning ahead, but they do not have a direct connection to the motor system.

Earlier research by Ungerleider & Mishkin argued for two separate pathways within the cortex. The dorsal visual pathway leads to the posterior parietal region, while the ventral visual pathway leads to the inferior temporal region. The authors relate these two basic streams to the concepts of vision for action and vision for perception respectively. Studies have shown that damage to the dorsal stream results in deficits in actions such as reaching, while the ability to distinguish perceived visual images remains intact. Other studies have shown that damage to the ventral stream creates difficulties with recognising objects, but does not impair vision-based actions such as grasping objects. An interesting study shows that attempts of patients with dorsal damage to point to images actually improved, if they delayed pointing until after the image had been removed. It was surmised that with the image gone, patients started to rely on a memory based on the intact ventral stream.

Neurons in the primary visual cortex fire in response to the position and orientation of particular edges, colours or directions of movement. Beyond the primary visual cortex, neurons code for more complicated features, and in the inferior temporal cortex they can code for something as specific as faces or hands. However, while neurons may respond only to quite specific features, they can respond to these with respect to a variety of viewpoints or lighting conditions.

By contrast neurons in the dorsal stream usually fire only when the subject responds to the visual signal, such as when they reach out to grasp an object. The ventral stream neurons appear to be moving the signals towards perception, whereas the dorsal stream neurons are moving signals towards producing action. The visuomotor areas of the parietal cortex are closely linked to the motor cortex. The authors suggest that the dorsal stream may also be responsible for shifts in attention at least those made by the eyes. The ventral stream has no direct connections to the motor region, but has close connections with regions related to memory and emotions, and providing information as to the function of objects.

Even within the ventral/perception stream there are separate visual modules. Damage related to any one module can result in localised deficits such as recognition of faces or of landmarks in the spatial environment. A cluster of visual areas in the inferior temporal lobe are responsible for most of these modules. This aspect of the research looks to point to the importance of a changing population of a small number of neurons or even single neurons in producing consciousness.

BLINDSIGHT

With the phenomenon of 'blindsight' patients have damage in the primary visual cortex V1. If V1 is not functioning, the relevant visual cells in the inferior temporal cortex remain silent regardless of what is presented to the eye. However, Larry Weiskrantz (25.), an Oxford neuropsychologist, showed that patients with this conditions could nonetheless move their gaze towards objects that they could not consciously see, and later studies showed they could scale their grip, and rotate their wrist to grasp such objects.

These blindsight abilities are mainly visuomotor. It is suggested that in these cases, signals from the eyes could go direct to the superior collicus, a midbrain structure that predates the evolution of the cortex, and thence to the dorsal stream. It is suggested that while the ventral stream depends entirely on activity in V1, there may be an alternative route for the dorsal stream. Studies have shown the dorsal stream to be active even when V1 is inactive. The patient discussed earlier is considered to be similar to blindsight patients, although in her case V1 is still active, which accounts for her still having conscious vision, albeit with impairments.

In this research visual perception provided by the ventral stream is seen as allowing us to plan and envisage the consequences of actions, and to file representations in long-term memory for future use. Motor control on the other hand requires immediate accurate information using a metric that correlates with the external world. In perception the metric is relative rather than absolute. Thus in a picture or film the actual size of the image doesn't matter, and we judge scale by the relative size of objects such as people or buildings. Thus the computation used for the absolute external accuracy of the motor system needs to be different from the relative computation of visual perception.

THE GAMMA SYNCHRONY

Probably the most established correlate of consciousness is the global gamma synchrony (53. New Horizons in the Neuroscience of Consciousness, 2010) (54. Dynamic Coordination in the Brain, 2010). Here it is important to repeat the distinction between correlation and identity. The fact that the gamma synchrony is correlated to consciousness, and that we understand how the synchrony arises does not of itself mean that we have explained consciousness. However, it seems reasonable to think that exploration of the gamma synchrony and its role might lead us towards an understanding of consciousness

THE BINDING PROBLEM

One important problem in consciousness studies is the so-called binding problem, as to why processing in spatially separated parts of the brain and in different modalities is experienced as a single unified consciousness. One the one hand, there is the unity of consciousness and on the other, the fact that the brain comprises a number of specialised although connected processing areas. In order for consciousness to become unified, it has to overcome the problem of being represented in different modalities. Further it is generally agreed that there is no single central processing area in the brain. Only a small part of the brain's total processing is conscious, and also much of the brain supports both conscious and unconscious processing. The studies of the gamma synchrony discussed below accord with the idea that it is this synchrony that creates the unity of consciousness.

The processing of neurons uses a fluctuation in electrical potential referred to as firing, spiking or axon potentials. The electrical potentials in individual neurons reach an axon terminal, which releases a neurotransmitter to a receptor on a neighbouring neuron. Axon spiking oscillates at particular frequencies. The gamma frequency of about 30-90 Hz, but mostly in the lower half of this region, is the most important frequency so far as consciousness is concerned. Numbers of neuronal assemblies can become synchronised to oscillate at this frequency.

Studies suggest that local gamma processing is unconscious, whereas large-scale activity, referred to as global, such as reciprocal signalling between spatially separate neural assemblies is a correlate of consciousness. Research indicates a close correlation between global gamma synchronisation and conscious processing (Lucia Melloni et al, 2007) (26.). Activity related to conscious responses is more synchronised, but not more vigorous. In human subjects, conscious processing has been related to phase-locked gamma oscillations in widely distributed cortical areas, whereas unconscious processing produces only local gamma activity.

This is argued to be a so-called 'small worlds' system, where there is a coexistence between local and long range networks. In the brain it is suggested that the local networks are between neurons only a few hundred micrometers apart within layers of the cortex, while the long distance networks run mainly through the white matter, and link spatially separated areas of the cortex. It is these latter that can establish a global synchrony that is correlated to consciousness.

EXPERIMENTATION

Melloni et al suggest that masking is a good way of studying consciousness, because this allows the same stimuli to be either conscious or unconscious. In a study run by the authors, words could be perceived in some trials but not in others. Local synchronisation was similar in both cases, but with consciously perceived words there was a burst of long-distance gamma synchrony between the occipital, parietal and frontal cortices. Also subsequent to this burst, there was activity that could have indicated a transfer of information to working memory, while an increase in frontal theta oscillations may have indicated material being held in working memory. Words processed at the unconscious level could lead to an increase in power in the gamma frequency range, but only conscious stimuli produced increases in long-distance synchronisation. This, plus possibly the theta oscillation, looks to be a requirement for consciousness.

In another study, long distance synchronisation in the beta as well as the gamma range was observed. Recent studies suggest a nesting of different frequencies of theta and gamma oscillations when there is conscious processing. Therefore long distance synchronisation looks to be a requirement for consciousness, and conscious stimuli are seen to be associated with phase-locking of gamma oscillations across spatially distributed regions of the cortex, and also with increases in synchrony without increases in the rate of neuronal firing rate.

A further study (Singer, Wolf: 2010) (27.) discusses the rhythmic modulation of neuronal activity. During processing in the cortex, the brain increasingly selects for the relationship between objects. This involves interactions between different parts of the cortex. There is a requirement to cope with the ambiguity of the external world. The environment may contain objects with contours that overlap, or are partly hidden, and these conflicting signals have to be resolved in the cortex. Further to this, some objects are encoded in different sensory modalities. Evidence suggests that this process involves not only individual neurons but also assemblies of neurons (Singer, 1999, Tsunoda et al, 2001) (28. & 29). The possible conjunctions in perception are too large to be dealt with by individual neurons, but instead utilise assemblies of neurons, with each neuron relating to particular aspects of the unified consciousness.

There appear to be two stages to this process. There is a signal to indicate that certain features are present. This operates on a 'rate code' basis, where a higher discharge frequency codes for a greater probability of a particular feature being present. The cortex is organised into neuronal columns extending vertically through the layers of the cortex. Synchronisation is related to connections linking these neuronal columns, which are thought to encode for linked features.

The inferior temporal cortex is regarded as the likely site for the production of visual objects, and object-related assemblies are associated with synchronisation. Oscillations are driven by inhibitory neurons through both synapses and gap junctions (Kopell et al, 2000, Whittington et al, 2001) (30 & 31). Inhibitory inputs to pyramidal cells favour discharges at depolarising peaks, and this allows synchrony in firing. Locally synchronised oscillations can become phase-locked with others that are spatially separated.

Synchronisation also allows better control of interactions between neurons. Excitatory inputs are effective if they arrive at the depolarising slope of an oscillation cycle and ineffective at other times. This means that groups of neurons that oscillate in synchrony will be able to signal to one another, and groups that are out of synchrony will be ignored. This mechanism can function both within neural assemblies or between separated assemblies. The frequency and phase of oscillation can alter so as to influence signalling.

FACIAL RECOGNITION

In one study, neurons responding to eyes, noses and faces were shown to synchronise to recognise a face. If the individual components were scrambled into a non-face arrangement then synchrony did not arise. However, the scrambling into non-face did not alter the discharge rate, only the synchrony. Focus of attention on objects also caused increased synchrony in the beta and gamma bands. Here again synchronisation does not necessarily relate to increased discharge rates.

CONVERGENCE ON SPECIALISED NEURONS

A further point of interest is the relationship between the global gamma synchrony and consciousness-related firing in single neurons. There is a tension at present between evidence relating subjective perception to the activity of large neuronal assemblies linked by the global gamma synchrony, and other studies relating it to the activity of much smaller numbers of neurons. Since the correlations with consciousness appear strong in both cases, it seems likely that consciousness will be found to involve both types of process. With studies related to small numbers of neurons, it is shown that neurons are selective for particular images or categories of image, and that most neurons will be inactive in relation to most objects.

FACE RECOGNITION AND LOCALISED HOT SPOTS

A recent study by Rafael Malach (32.-35.) indicates that while perception involves widespread cortical processing, the emergence of an actual percept sometimes involves only a small number of localised hot spots, in which there is intense and persistent gamma activity. Malach used the area of face recognition to clarify this concept. Studies indicate the existence of so-called totem cells (a reference to totem poles with carved faces) that are able to recognise a number of faces. The hot spots are suggested to involve intense activity between several of these totem neurons resulting in a sort of vote. If the same face is recognised by a majority or most of the neurons, the face is consciously recognised. The presumption seems to be that this would apply for most forms of perception and not just face recognition.

Malach's studies hint at important possibilities. Firstly, they raise the game for the individual neuron, from a simple switch, to something probably involved in more sophisticated processing. If we accept a conscious processing role for the individual neuron, it puts a different light on the global gamma synchrony, as a possibly classical structure that simply coordinates the activity of a number of hot-spot neurons, in order to produce the unity of consciousness. Thus in Malach's example, face-recognition is not the end of the problem, because we do not usually perceive faces in isolation but as part of an environment. This suggests that the gamma synchrony could ensure that face recognition is coordinated with other hot-spot neurons that recognise clothing, furniture, a room or a surrounding landscape.

'ALL OR NOTHING' NEURONS

A further study also involving Malach looked at the response of single neurons in the medial temporal lobe, while subjects looked at pictures of familiar faces or landmarks. The response of the neurons studied correlated with conscious perceptions reported by the subjects of the study. Visual perception is processed by the ventral visual pathway, which goes from the primary visual cortex to the medial temporal lobe. Recent studies have shown that neurons in the medial temporal lobe fire selectively to images of individual people. In some trials, the duration of stimuli was right on the boundary of the time needed for conscious recognition of an object, so that it was possible to compare the behaviour of the neurons when an object was recognised and not recognised by the subject.

One finding of this study was the 'all-or-nothing' nature of the neuronal response. There was no spectrum involved. Either the neuron fired strongly, in correlation with the subject reporting recognition, or there was very little activity. The responses were not correlated with the duration of the stimuli, because the responses of the neurons lasted considerable longer than the stimuli.

In one trial, a single neuron was shown to respond selectively to a picture of the subject's brother, but not to other people well known to the subject. Particularly noted is the marked difference in the firing of the neuron when the subject's brother was recognised and not recognised. The stimulus duration of 33 ms meant that half the time the image was recognised, and half the time not recognised. The neuron was nearly silent when the image was not recognised, but fired at nearly 50 Hz when there was conscious recognition, indicating an 'all-or-nothing' response from the neuron, correlated to subjective report of recognition. The response exceeded the duration of the stimulus, and it was shown that the range of signal duration had little influence on the neuron's response.

In another test, a single neuron went from baseline to 10 spikes per second when the subject recognised a picture of the World Trade Centre, but showed little response to all other images that were presented. Again the neuron fired in an 'all-or-nothing' fashion, depending on whether there was conscious recognition. In five trials not resulting in conscious recognition, this neuron did not fire a single spike. In yet another trial, the firing of a single neuron jumped from 0.05 Hz to 50 Hz when the subject reported recognition of an individual.

The overall conclusion from these trials is that there is a significant relationship between the firing of neurons in the medial temporal region and the conscious perceptions of subjects. Further to this, the activity of the neurons lasted for substantially longer than the stimuli, and had only a marginal correlation with the stimuli. In particular, it is noted that with stimuli, at a duration where exactly the same image was recognised in some cases, but not in others, there was an entirely different (all-or-nothing) response from the neuron, according to whether or not the subject consciously recognised the image. Other neurons near to the medial temporal neurons studied were shown to respond to different stimuli from those that activated the studied neurons. These findings are stated to agree with earlier single-cell studies, including studies involving the inferior temporal cortex and the superior temporal sulcus.

This study serves to refute one of the popular arguments of twentieth century consciousness studies to the effect that consciousness was 'just what it was like to have a brain or neural processing'. The study demonstrates that neural processing is completely distinct for exactly the same signal, with a duration that placed it on the boundary of being consciously recognised or not recognised, produced almost no response, if it was not consciously recognised, but a vigorous response if it was consciously recognised.

NEURONAL SPIKING, CONSCIOUSNESS AND THE GAMMA

A study by Rafael Malach et al shows a correlation between consciousness and a jump from baseline to 50 Hz spiking in single neurons. Rather similar experiments show a correlation between global gamma synchrony and conscious experience. The problem here is to discover the link, if any, between these two correlations. The authors ask to what extent the spiking activity of individual neurons is related to the gamma local field potential. Earlier studies had shown a confusing variation in the degree of correlation between neuronal spiking and gamma activity, with some studies showing a strong correlation and others showing only a weak correlation. The authors here think that they have a resolution to the arguments that have arisen around this confusing data.

Their study demonstrates that most of the variability in the data can be explained in terms of whether or not the activity of individual neurons is correlated to the activity of neighbouring neurons. A relationship with the gamma synchrony is apparent where there is correlated activity in neighbouring neurons. The link between individual neurons that are associated with other active neurons and the gamma synchrony is apparent, both when the brain is receiving sensory stimulation, and when activity is more introspective.

The gamma synchrony is considered to arise from the dendritic activity of a large number of neurons over an extensive area of the cortex. This study shows that the relation between the activity of individual neurons and gamma correlates with the extent to which the activity of the neuron is linked to the firing rate of its neighbouring neurons. This establishes a relationship between gamma activity and a large number of individual neurons distributed over a region of the cortex.

In this study discussed here, subjects watched a film. During this, scanning showed a high correlation between the spiking of individual neurons and gamma activity that arose at the same time. But this did not happen in all cases. It was found that the main factor relating to whether or not neuronal spiking related to gamma activity was the degree of correlation in spiking between neighbouring neurons. This study was based on recording the activity of several individual neurons. It was shown that the correlation between the spiking of the individual neuron and gamma synchrony could be predicted from the level of correlations between the activity of neighbouring neurons. When neurons were not correlated with their neighbours gamma activity was at a low level.

CONSCIOUSNESS AND THE SENSORY CORTEX

Rafael Malach again argues that, at least in some cases, conscious perception does not require any form of 'observer' in the prefrontal area, but needs only activation in the sensory cortex. This claim is based on fMRI studies performed by Malach and colleagues. In one study where subjects had their brains scanned while watching a film, there was a wide spread activation of the sensory cortex in the rear of the brain, coinciding with relatively little activity in the frontal areas, where a significant degree of inhibition was apparent. Malach pointed out that the use of a film contrasted with the more normal brain scanning procedure in which stationary objects are presented in isolation without being embedded in a background and without other modalities such as sound.

Malach argued that the use of a film contrasted with the more normal brain scanning procedure in which stationary objects are presented in isolation without being embedded in a background and without other modalities such as sound. With subjects watching a film there was synchronisation across 30% of the cortical surface. This synchronisation extended beyond both visual and auditory sensory cortices into association and limbic areas. Emotional scenes in the film were correlated with widespread cortical activation. The study appears interesting in terms of the ability to synchronise more than one sensory modality plus the emotional areas of the brain. It was further shown that the more engaging the film, the less activity there was in the frontal areas.

Malach suggests that the role of the frontal areas is not to create perceptual consciousness but to deliberate on the significance of the sensory experience and to make it reportable. When introspective or deliberative activity is in process, it is accepted that both sensory and prefrontal areas may be activated. If we accept this approach it becomes impossible to explain consciousness entirely in terms of the self, and the easy let out of deconstructing the self, and then claiming to have explained consciousness is closed off.

One study (Hasson et al, 2004) also scanned brain activation in subjects viewing a film. In general, the rear part of the brain, which is orientated towards the external environment, demonstrated widespread activation. In contrast, the front of the brain and some areas of the rear brain showed little activation. These less active areas are referred to as the 'intrinsic system' that deals with introspection, and the 'first person' or 'self' aspects of the mind. Reportability is presumed to arise in this part of the brain. This network shows a major reduction in activation at the times that perception is most absorbing. This observation is exactly the reverse of any notion that perception and reporting should work in tandem.

Malach suggests that conceptually, there could be an axis running from, firstly, introspective activity in the prefrontal, through, secondly, attention to external world material such a film, which can activate much of the sensory cortex, while inhibiting prefrontal activity, to thirdly and finally experiences such as Zen meditation, which can be seen as pure perception without any residual awareness of the self. This type of pure perceptual/absence of self experience is reported as being associated with other forms of altered states of consciousness.

On a more everyday level, Malach suggests that when subjects are sufficiently absorbed by their sensory perceptions, they 'lose themselves' in the sense of not having any introspection about what they are perceiving. A typical example is an interesting film in which the viewer is absorbed by the drama and suspends any personal introspection or attempts to report what they are experiencing.

It is also stressed here that consciousness arises of its own accord in the sensory cortex, without being dependent on frontal cortices supposed to be related to the sense of self. This looks to undermine attempts to dismiss the problem of consciousness by conflating it with the self, and then after that deconstructing the self. On the other hand, it would probably be going too far in the other direction to say that consciousness does not arise at all in the frontal areas. In particular some activity in the orbitofrontal cortex can be correlated to conscious perception rather than the strength of the signal, in much the same way that Malach has indicated occurs with visual perceptions. Dorsolateral activity also appears to only arise when there is a correlation with reportable conscious experience.

Malach speculates that these experimental findings support the idea that subjective experience arises in the areas where sensory processing occurs, rather than having to be referred on to any type of higher-order read out or some form of separate 'self'. In this view, sensory perceptions are seen as arising in a group of neurons. Studies show that high neuronal firing rates over an extended duration and dense local connectivities of neurons are associated with consciousness. Malach thinks that studies of brain processing can differentiate conscious perception from the process of reporting the perceptions and the self, but can be handled by groups of neurons, within which individual neurons provide the perceptual read-out or subjective experience.

He also argues that this supports the view that consciousness arises in each of a number of single neurons in a network, rather than having to refer to some higher structure. The perception arises when all the neurons in a particular network are informed about the state of the others in the network. Thus the perception is suggested to be both assembled by, and read-out or subjectively experienced by the same set of neurons. Each active neuron is suggested to be involved in both creating and experiencing the perception.

This view of conscious perception has some important implications for consciousness theory as a whole. In the first place, it makes it possible to consider looking for the process by which consciousness arises in individual neurons rather than brain wide assemblies. This is more easily consistent with the recent findings that quantum coherence and possibly entanglement is functional in individual living cells. A further point is that the idea of consciousness in neurons or small high density areas undermines the attempt by some consciousness theorists to try and conflate consciousness with the self and self-consciousness, and then claim that a deconstruction of the self has explained consciousness.

AMBIGUOUS IMAGES

Similar evidence emerges from studies of the well-known Rubin ambiguous vase-face illusion. High fMRI activity correlates with the emergence of a face perception, although this emergence into consciousness does not involve any alteration in the external signal (Hasson et al, 2001). This is another demonstration that brain activity can correlate to conscious perceptions rather than the nature of external signals.

The authors consider that consciousness is correlated with non-linear increases in neural activity, here described as 'neuronal explosions' and occurring in sensory areas. Other fMRI studies have distinguished two types of fMRI reading. Sensory activity is marked by rapid but short bursts of neuronal firing, while rest activity in neurons involves slow, low amplitude activity.

FURTHER SELECTIVE RESPONSE STUDIES

Quiroga, Q. et al (2008) (36.) emphasise that studies over the last decade have shown that some neurons in the medial temporal lobe respond selectively to complex visual stimuli. The studies suggest a hierarchical organisation along the ventral visual pathway. Neurons in V1 code for basic visual features, whereas at the stage of the inferior temporal cortex neurons can code selectively for complex shapes or even faces.

The inferior temporal cortex projects to the medial temporal cortex where neurons are found to be selectively responsive to categories such as animals, faces and houses, as well as the degree of novelty of images. Activity in the medial temporal lobe is thought to be linked to creating memories rather than actual recognition, a process that seems to be more closely linked to the inferior temporal lobe.

In a study by the authors, a hippocampal neuron fired in response to the image of a particular actor. Recording of the activity of a handful of neurons could be used to predict which of a number of images a subject was viewing at an accuracy far above chance. About 40% of medial temporal lobe neurons were found to be selective in this way, although some could fire selectively in response to more than one image. However, when this was case the images were often connected, such as two actresses in the same soap opera, or two famous towers in Europe. In fact it is estimated that selectively responding cells would respond to between 50 and 150 images.

The authors are not trying to revive the idea of the 'grandmother cell' where one and only one neuron could respond to a particular image, for instance the image of the subject's grandmother. Rather than that, the authors have estimated that out of one billion cells in the medial temporal lobe, two million could be responsive to specific percepts. These cells respond to percepts that are built up in the ventral pathway rather than detailed information falling on the retina.

DISTINCTION BETWEEN PHYSICAL INPUT AND CONSCIOUS PERCEPTS

Kreiman, Fried & Koch (2002) (37.) demonstrated that the same environmental input to the retina can give rise to two quite different conscious visual percepts. In this study, the responses of individual neurons were recorded. Two-thirds of the visually selective medial temporal lobe neurons recorded showed changes in activity that correlated with the shifts in what was subjectively perceived, rather than the retinal input. Flash suppression is an experimental technique by which an image is sent to one eye and then a different image to the other eye. The newer image will suppress the first input. Neurons that select for the initial input and not the input to the second eye will be inactive when the first input is suppressed in this way, although the first image is still physically present on the retina.

In visual illusions such as the Necker cube, the same retinal input can produce two different subjective perceptions. There is a distinction here between what happens in the primary visual cortex and in the later visual areas. Activity in the primary areas correlates to the retinal input, rather than any subjective perception. In this study performed in the US in 2002, a neuron in a subject's amygdala responded selectively to the image of President Clinton, while failing to respond to 49 other test images presented. In the case of Clinton's image, the neuron's firing rate jumped from a baseline of 2.8 spikes a second to 15.1 spikes per second. However, the neuron did not react when the initial image of Clinton was suppressed by an image for which the neuron was not selective. Another amygdala neuron increased its firing in response to some faces, but was inactive when an image it didn't select for was flashed to the other eye. A neuron in the medial temporal lobe increased its firing in response to pictures of spatial lay outs and not to other stimuli. Here again the activity did not occur when a different image was flashed to the other eye. In all these cases, the physical input to the first eye was continuing, but was not getting into conscious perception.

Out of 428 neurons studied in the medial temporal lobe, 44 responded selectively to particular categories and 32 to specific images. None of these neurons were active when the images or categories they were selective for were part of the input on their retina, but were suppressed from subjective experience by a second image to the other eye. However, they could be active when both images were present, but the image they selected for was dominant. In the experimental subjects two out of three medial temporal lobe neurons changed their firing in line with subjective perceptions, but activity did not change if an input was present on the retina but not subjectively experiences because of retinal input to the other eye.

This study could be seen as laying to rest two favourite ideas of twentieth century consciousness studies. The first was the idea that consciousness was non-physical. This approach is not really coherent within a scientific paradigm in any case, but experiments now demonstrate a correlation between subjective perceptions and physical levels of activity in individual neurons. Similarly, the mind-brain identity concept seemed to propose that in some mysterious way consciousness was identical to the whole operation of the brain, whereas this and other experiments clearly relate consciousness to the activity of individual neurons and specific neuronal assemblies, albeit both the neurons and assemblies involved are constantly changing.

OBJECT RECOGNITION

Kalanit Grill-Spector discusses studies with fMRI that have shown that activation in particular brain regions correlates with the recognition of objects and also of faces. Some regions are involved in both face and object recognition. Object recognition occurs in a number of regions in the occipital and temporal cortex collectively referred to as the lateral occipital complex (LOC). These regions respond more strongly when the subjects are viewing objects. The involvement of LOC is thought to be subsequent to the early visual areas (V1-V4) and in the ventral stream, responding selectively to objects and shapes and showing less response to contrasts and positions. There are object-selective regions in the dorsal stream, but these do not correlate with object perception, and are suggested to be involved with guiding action towards objects. The LOC is responsive to objects without reference to how the object is defined, i.e. it does not differentiate between a photograph and a silhouette. The LOC responds to shapes rather than surfaces, and it responds even if part of the shape is missing.

It is suggested that a pooled response across a population of neurons allows a response to objects that does not vary according to the position of an object. This could be taken to indicate a role for individual neurons. Each neuron's response varies according to the position of the object. It appears that for any given position in the visual field each neuron's response is greater for one object than for all other objects presented.

Apart from the LOC other regions in the ventral stream have been shown to respond more to particular categories of object. One region showed more response to letters, several foci responded more to faces than objects, including the fusiform face area, while other areas responded more to buildings and places than to faces.

Nancy Kanwisher et al have suggested that the ventral temporal cortex contains modules for the recognition of particular categories such as faces, places or parts of the body. However, it is suggested that the processing of faces is extended to a more sophisticated level, given the requirements for social interaction. There may be a distinction between processing to recognise individual faces and processing to recognise categories, such as horses or dogs as a category. However, it is suggested very expert recognition of categories, such as ornithologists recognising a bird may involve a process similar to face recognition. Rafael Malach et al suggest that category recognition may respond more to peripheral input, while face and letter recognition depends more on central stimuli.

GAMMA, NEURONS & CONSCIOUSNESS

This could suggest that the conscious response in a single cell is linked to or dependent on global gamma synchrony. However, it would appear not necessary for the whole collection of neuronal assemblies to come into consciousness, but only for the synchrony to trigger consciousness in the individual neuron. This might make it possible to invert Hameroff's proposal for quantum coherence in neurons to drive consciousness in the gamma synchrony. The opposite case of the synchrony triggering consciousness in single neurons would be more compatible with the type of quantum coherence that is functional in photosynthetic organism.

This tends to look like pieces of a jigsaw puzzle, and unfortunately one that we may not get much help in assembling. We know that the global gamma synchrony correlates to consciousness. We know that a jump to 50 Hz spiking in individual neurons correlates to consciousness. We also now know that the spiking in the individual neurons correlates to gamma if the spiking of the individuals correlates to their neighbours.

From the point of view of recent findings relative to quantum coherence in organic matter, it has become most plausible to think in terms of consciousness arising within individual neurons, but the road there may involve feed forward and feedback, as is often the case in brain processing. Processing in one neuron as a result of external signals may set off other neighbouring neurons, which ultimately broaden into a neuronal assembly oscillating as a local gamma synchrony. Longer range signals to other neuronal assemblies would set up global gamma synchrony. It might only be at that point that signals went back to individual neurons triggering quantum coherent activity within the neuron. This might account for the 500ms time lag for signals to come into consciousness (the Libet half second), while at the same time being compatible with the femto and pico second timescales of functional quantum activity in biological systems. Very speculative, but perhaps this at least provides a starting point or rational framework for thinking about the consciousness problem.

THE EMOTIONAL BRAIN

In recent years, the most important neuroscientific research has arguably involved the role of emotion or emotional evaluation in the brain. This was a previously very neglected area due to various biases and misconceptions in twentieth century neuroscience. Our attention is here focused on how the orbitofrontal cortex assigns reward/punisher values to representations projected from other cortices, and how the basal ganglia integrate these subjectively-based values with inputs from the other parts of the cortex and the limbic system.

SUBJECTIVE EMOTION, CHOICE &
A COMMON NEURAL CURRENCY

We are conscious of emotions, and they allow us to assess the reward values of actions. Without the emotion-based assessment of rewards, rational processing is not by itself adequate to deliver normal behaviour. While reasoning can be seen as working with or without consciousness, the subjective experience of emotion is closely entwined with the subjective assessment of current or future rewards. In fact, this ability to have a subjective preference or choice can be argues to be the real distinction between conscious and non-conscious systems, the difference between an automated one-to-one response and the conscious but unpredictable preference of one thing over the other.

Emotion, anticipation of rewards and enjoyment of the same are all here seen to be based on subjective experience, and the key importance of these factors for behaviour suggests that subjective emotion is a common neural currency underlying the determination of behaviour. It is hard to distinguish a purely algorithmic basis for this processing, since the weighting of two subjective experiences seems to require the injection of initially arbitrary weights suggesting a non-computable or non-algorithmic element.

REWARDS AND PUNISHERS

Modern descriptions of emotional processing in the brain revolve round a framework of 'rewards' and 'punishers', together referred to as 'reinforcers', with subjects working to gain rewards and to avoid punishers. Some stimuli are primary reinforcers, so-called because they do not have to be learned. Other stimuli are initially neutral, but become secondary reinforcers, because through learning, they become associated with pleasant and unpleasant stimuli.

Reward assessment is argued to be implemented in the orbitofrontal region of the prefrontal cortex and in the amygdala, a part of the subcortical limbic system. Emotions are thus viewed as states produced by reinforcers. The amygdala, the orbitofrontal and the cingulate cortex are seen as the brain areas most involved with emotions. Emotional states are usually initiated by reinforcing stimuli present in the external environment. The decoding of a stimulus in the orbitofrontal and amygdala is needed to determine which emotion will be felt.

NEUTRAL REPRESENTATIONS

In respect of emotions, the brain is envisaged as functioning in two stages. To take the best known example of the visual system, input from the eyes is processed in the rear (occipital) area of the brain, and then progressively assembled into a conscious image arising in the inferior temporal cortex. At this stage, these representations are neutral in terms of reward value. Thus visual representations in the inferior temporal, or analogous touch representations in the somatosensory cortex, are shown to be neutral in terms of reward value, until they have been projected to the amygdala and the orbitofrontal. The brain is organised first to process a stimulus to the object level, and only after that to access its reward value. Thus reward/punisher values are learned, in respect to perceived objects produced by the later stages of processing, rather than the pixels and edges produced by the earlier stages of processing.

ORBITOFRONTAL CORTEX—SUBJECTIVE EXPERIENCE OVER STRENGTH OF SIGNAL

The orbitofrontal region of the prefrontal cortex is seen as the most important region for determining the value of rewards or punishers (55. Zald & Scott, 2006). Objects are first represented in the visual, somatosensory and other areas of the cortex, without having any aspect of reward value. This only arises in the orbitofrontal and the amygdala. Studies show that orbitofrontal activity correlates to the subjective pleasantness of sensory inputs, rather than the actual strength of the signal. The orbitofrontal projects to the basal ganglia, which appear to integrate a variety of cortical and limbic inputs in order to drive behaviour. Thus subjective emotional assessment occurring mainly in the orbitofrontal would appear to play an important part in determining behaviour.

The orbitofrontal cortex receives input from the visual, auditory, somatosensory and other association cortex, allowing it to sample the entire sensory range, and to integrate this into an assessment of reward values. In the orbitofrontal, some neurons are specialised in dealing with primary reinforcers such as pain, while others are specialised in dealing with secondary reinforcers. Orbitofrontal neurons can reflect relative preferences for different stimuli. The subjective experience of one signal can be altered by another from a different modality. The impact of words can influence the subjective impression of an odour, and colours can also influence the perception of odour. It has also been shown that the subjective quality of, for instance odours, can be altered by the top-down modulatory impact of words, while colour is thought to influence olfactory judgement. There is seen to be a triangular system involving association cortex, amygdala and orbitofrontal.

EXPERIMENTATION: CORRELATION WITH SUBJECTIVE EXPERIENCE

An important study looked at the activation produced by the touch of a piece of velvet and a touch of a piece of wood in the somatosensory cortex, and the activation of the orbitofrontal produced by the same touches (38-40. Rolls, Edmund). This trial compared the pressure of a piece of wood with the perceived pleasant pressure of a piece of velvet. It was demonstrated that the pressure of the wood produced a higher level of activity in the somatosensory cortex than the pressure of a piece of velvet. However, in the orbitofrontal the same pressure from velvet produced a higher level of activation, with the difference between velvet and wood being correlated to the different subjective appreciation of the two pressures.

The less intense but reward-value positive stimuli, produced more activation in the orbitofrontal than a more intense but reward-value neutral stimulus. Similarly a reward value negative stimulus also produced more activation in the orbitofrontal than a neutral stimuli that was registered as stronger by the somatosensory cortex. Researchers are clear in their conclusion that the orbitofrontal registers emotionally positive or negative aspects of an input, rather than any other aspects such as intensity of signal. Thus the subjective pleasantness of the velvet touch relates directly to the activation level of the orbitofrontal cortex, demonstrating a connection between subjective appreciation and the core mechanisms for decision taking and behaviour. The orbitofrontal and to a lesser extent the anterior cingulate cortex are seen here as being adaptive in registering the emotional or reward value aspects of the initially reward-neutral somatosensory stimulation.

Studies suggest that the orbitofrontal deals with a variety of types of reward values. It has been suggested that the brain has a common neural currency for comparing very different reward values. Apart from the velvet/wood study, other studies show that the level of orbitofrontal activity correlates to the subjective pleasure of the sensation, rather than the strength of the signal being received. Activation in response to taste is seen to be in proportion to the subjective pleasantness of the taste, and in responding to faces, activity increases in line with the subjective attractiveness of the face. With taste, the orbitofrontal can represent the reward value of a particular taste, and this activation relates to subjective pleasantness. In humans the subjectively reported pleasantness of food is represented in the orbitofrontal. Studies of taste in particular are seen as evidence that aspects of emotion are represented in the orbitofrontal. With faces, the activation of the orbitofrontal has been found to correlate to the subjective attractiveness of a face. This subjective ability enables flexibility in behaviour. If there is a choice of carrots or apples, carrots might be preferred and the top preference signal in the brain would correlate to carrots. However, if the range of choice was subsequently expanded to include bananas, the top preference signal could switch to bananas. This reaction looks to require some form of preferred qualia, referring to a previous subjective experience of bananas.

DIFFERENT TYPES OF REWARD—MONEY V. SEX

One study attempted to compare the brain's processing of monetary rewards, with its processing of rewards in terms of erotic images. Monetary rewards were shown to use the anterior lateral region of the orbitofrontal cortex, while erotic images activated the posterior part of the lateral orbitofrontal cortex and also the medial orbitofrontal cortex. Brain activity in these orbitofrontal regions increased with the intensity of reward, but only for types of reward in which those areas were specialised. By contrast, activity increased for both monetary and erotic rewards in the ventral striatum, the anterior cingulate cortex, the anterior insula and the midbrain.

Other studies using rewards such as pleasant tastes have suggested a similar distinction between the posterior and anterior regions of the orbitofrontal. The bilateral amygdala was the only subcortical area to be activated in reward assessment and it was only activated by primary rewards such as erotic images and not by abstract rewards such as money. This area is more strongly connected to the posterior and medial orbitofrontal than to the anterior orbitofrontal.

One distinction that is argued to emerge is between immediate reward, and the more abstract quality of a monetary reward that can only be enjoyed over time. The authors argue that studies suggest that it is not the actual delay in benefiting from the monetary reward, but its abstract nature that leads to it being processed in a different area. It was also found that patients with damage to the anterior orbitofrontal have difficulty with assessing indirect consequences as distinct from immediate consequences.

ADAPTIVE ADVANTAGE OF FLEXIBILITY AND RESPONSE TO CHANGE

The adaptive advantage of the emotional system is that responses to situations do not have to be pre-specified by the genes, but can be learned from experience. If evolution had attempted to specify fixed responses for every possible stimuli, there would have been an unmanageable explosion of programmes. The reinforcer defines a particular goal, but does not specify any particular action.

The orbitofrontal is also suggested to be involved in amending responses to stimuli that used to be associated with rewards, but are no longer linked to these. Three groups of neurons in the orbitofrontal provide computation, as to whether reinforcements formerly associated with particular stimuli are still being obtained. These neurons are involved in altering behavioural responses. The orbitofrontal computes mismatches between stimuli that are expected and stimuli that are obtained and changes reward representations in accord with this. This rapid reversal of response carries through from the orbitofrontal to the basal ganglia. Damage to the orbitofrontal impairs the ability to respond to such changes, and is associated with irresponsible and impulsive behaviour, and difficulty in learning which stimuli are rewarding and which are not. Patients who have suffered damage to the orbitofrontal have difficulty in establishing new and more appropriate preferences, and in daily life they tend to manifest socially inappropriate behaviour. In particular there is greater difficulty in dealing with indirect or longer term consequences of actions than with direct and immediate consequences.

VISCERAL RESPONSES AND EMOTIONS

The orbitofrontal and amygdala act on the autonomic and the endocrine systems when stimuli appear to have significance in terms of emotion or danger. Visceral responses as a result of this signalling are fed back to the brain. Studies suggest that visceral responses are integrated into goal-directed behaviour via the ventromedial prefrontal cortex (VMPFC). The insula and the orbitofrontal are also thought likely to map visceral responses, with feedback from the viscera influencing reward assessment via levels of comfort or discomfort.

There is considerable support for the idea that the body is the basis of all emotion. However, this looks difficult to square with the actual structure and nature of brain processing. While the bodily responses can certainly be seen to play a role, it is hard to see why all visual, auditory inputs, and the results of cognitive processing should have to wait on the laborious responses of the viscera, especially as it is the reward assessment areas of the brain that signal the viscera in the first place. If bodily emotion were the whole story, the orbitofrontal and amygdala would seem to be in a state of suspended activity between sending a signal to the autonomic system and getting signals back from the viscera. Conventional thinking may have here been biased by the emphasis in experimentation on fear in animal subjects, where bodily reactions are pronounced, rather than more evaluative emotional activity emphasised above.

In the specific case of rapid phobic reactions in the amygdala, the idea fails completely. The more plausible view is that visceral responses are one aspect of many responses that are integrated in the orbitofrontal. It seems more likely that in line with most brain processes there is a complex feed forward and feedback between all parts of the system including the viscera and the orbitofrontal. The body-only theory seems to depend on a simple feed forward mechanism, which is alien to how brain processing works.

DORSOLATERAL PREFRONTAL

The orbitofrontal projects not only to the basal ganglia but also to the dorsolateral prefrontal, which is responsible for executive functions, planning many steps ahead to obtain rewards, and such decisions as deferring a short-term reward in favour of a higher value but longer-term reward. Where dorsolateral activity reflects preferences, it is found that the orbitofrontal has reflected them first, and these preferences have been projected from the orbitofrontal to the dorsolateral, where they can be utilised for planning or for deciding whether or not to defer short-term rewards. In these instances, the reward assessing functions of the orbitofrontal and the integrative role of the basal ganglia play an important role. It has been argued that that 'moral-based' knowledge generated by rewards and punishers cannot take place without the orbitofrontal. Ethically based rewards for good or appropriate behaviour that are decided on by the dorsolateral are seen to be influenced by processing of the orbitofrontal.

THE BASAL GANGLIA

In the basal ganglia, the emotional evaluation of the orbitofrontal is combined with inputs from other cortices and the limbic areas (41. Leonard Kosiol & Deborah Budding, 2010). The basal ganglia can be viewed as a sort of mixer tap for the wide spread of inputs from the cortex and limbic system, and as such select or gate for material processed by the cortex, including the orbitofrontal. The basal ganglia comprise a region of the brain with strong projections from most parts of the cortex and also the limbic system. Modern brain theory views the basal ganglia as important for the choice of behaviours and movements, both as regards activation and inhibition of these. The striatum, which includes the nucleus accumbens, is the largest component of the basal ganglia, receiving projections from much of the cortex, and also receiving dopamine projections from the midbrain. The basal ganglia are sensitive to the reward characteristics of the environment, and operate within a reward-driven system based on dopamine. The region is seen as integrating sensory input, generating motivation and also releasing motor output.

Incoming stimuli from the environment to the brain are always excitatory. The thalamus receives the incoming signals, and sends them forward to the cortex for processing. This is also primarily excitatory, as are further projections to the frontal cortex. The basal ganglia are seen as important for inhibition. Cortical-subcortical-cortical loops are widespread in the brain. In these loops, the cortical inputs are always excitatory, with the subcortical for the most part inhibitory. The subcortical areas are seen to project back to the cortex, and to modulate the cortical inputs. They are indicated to have a role in deciding what information is returned to the cortex. Each loop originates in a particular area of the cortex, such as the orbitofrontal and the anterior cingulate. Inhibitory output going back via the thalamus assists the focusing of attention and action. The basal ganglia gate or select for elements of the processed information used by the cortex. Novel problem solving requires interaction between the prefrontal cortex, other parts of the cortex and the basal ganglia.

STRIOSOMES, MATRISOMES & TANS (TONICALLY ACTIVE NEURONS)

Striosomes are the area of the basal ganglia involved in modulating emotional arousal. The basal ganglia includes the striatum, which contains neurochemically specialised sections called striosomes that receive inputs mainly from limbic system structures, such as the amygdala, and project to dopamine containing neurons. This is seen as giving them a role in dealing with the input of emotional arousal into the basal ganglia (Graybiel, 1995). Certain regions in the cortex, and notably areas involved with emotion such as the orbitofrontal cortex, the paralimbic regions and the amygdala, all project to the striosomes (Eblen & Graybiel, 1995). This is seen as constituting a limbic-basal ganglia circuit. The role of the striatum may be to balance out a variety of sometimes conflicting inputs from different parts of the prefrontal and the limbic areas, and to switch behaviour in response to these inputs.

In the mid 1990s researchers discovered specialised neurons referred to as tonically active neurons (TANs) that are situated where matrisomes and striosomes meet, and are therefore well placed to integrate emotional and rational input. Cortical areas involved with anticipation and planning project to areas in the striatum known as matrisomes. These are often found in close proximity to the striosomes. This is taken to suggest a link between the planning-related matrisomes and the limbic-related striosomes. TANS (tonically active neurons) are highly specialised cells located at striosome/matrisome borders, and therefore well placed to integrate emotional and rational input. TANS can be seen as a form of mixer tap for combining planning and emotional assessment inputs in the basal ganglia.

TANs respond strongly to reward-linked stimuli, and they also responded when a previously neutral stimuli becomes associated with a reward. TANs are thought to be involved in the development of habits, with particular environmental cues having emotional meaning, and producing particular behaviour. These cells have a distinct pattern of firing when rewards are delivered during behavioural conditioning (Asoki et al, 1995). It is suggested that changes in TAN activity could be a way of redirecting information flow in the striatum.

DOPAMINE

The neurotransmitter dopamine is involved in delivering the reward system for which the orbitofrontal and other areas act as a prediction. The largest concentrations of dopamine in the brain are found in the prefrontal cortex and the basal ganglia. The dopamine system is based in the ventral tegmenta area of the midbrain. The dopamine producing neurons in the mid brain appear to be influenced by the size and probability of rewards, presumably based on information from areas such as the orbitofrontal and the amygdala. Dopamine projections are mainly to the nucleus accumbens, the amygdala and the frontal cortex. This is the brains reward circuitry. The ventral striatum is highly active in anticipation of reward, and remains active during the reward. It is believed to modulate motivation, attention and cognition. Impairment of this area creates a wide range of problems. Within the striatum learning is influenced by dopamine acting on medium spiny neurons, reducing inhibition and releasing or increasing output of activity. By contrast, reduced levels of dopamine lead to increased inhibition and reduced activity.

REWARD/PLEASURE CENTRE—NUCLEUS ACCUMBENS

The nucleus accumbens is part of the ventral striatum and constitutes the reward/pleasure centre of the brain. The orbitofrontal and anterior cingulate both project to the nucleus accumbens. Dopamine-based activity in the nuclear accumbens is related to seeking reward and avoiding pain. Addictions are found to be related to a lack of natural activity in this area, with drugs of addiction working to enhance otherwise depressed activity. It has further been suggested that the use of neuromodulators by-passes the need to always rely on cognitive computation in the cortex. From the point of view of consciousness studies, it is apparent that these dopamine-rewards are registered in subjective consciousness, so as with the orbitofrontal there is again a weighting of different subjective impulses. The orbitofrontal would look to base its predictions on the previous subjective experience of the delivery of dopamine rewards.

FREEWILL

The nature of emotional evaluation in the brain discussed above leads on to the vexed question of free will. The area of conventional consciousness studies has been almost unanimous in rejecting the concept of freewill in favour of human behaviour being completely deterministic. We do not have to look very far to find the explanation for this counter-intuitive notion. Conventional thinking about consciousness is based on classical/macroscopic physics, which is entirely deterministic and has no place for anything outside a direct sequence of cause and effect. The high degree of confidence expressed in deterministic explanations rests on this assumption. Once we begin to think that classical physics might not have the full explanation for consciousness, the assurance of determinism looks to be shaky. This in itself may partly explain the furious resistance to the involvement of non-classical physics in brain processing.

Recent studies of the processing of emotion in the brain do not accord well with the deterministic thesis, albeit not many have yet come to terms with this. The workings of the emotional brain provide something that can only be experienced in terms of subjective scenarios, not apparently reducible to specific weightings or to algorithms that give precise and deterministic predictions. Another way of approaching the problem of deciding between two alternative courses of action is to look at what happens when we make a list of points in favour of both courses of action. While this may somewhat clarify the mind, we are still likely to find that something is missing, something which will ultimately need to be bridged by an emotional evaluation.

Whether such subjective based decisions or influences can be described as 'free' is hard to say. They certainly look to lie outside the classical-based neuroscience which is the usual diet provided for us, but whether they represent 'free' agency is another matter. As something not derived from algorithms, they can however be seen as deriving from the same fundamental level of the universe that can over some extended period of time give a pattern to the apparently randomly arising position of particles. It is perhaps beyond us at the moment to say what is that takes this sort of decision, but what they have in common with emotional evaluation is that they cannot be described by an algorithm. In the discussion below, we look at various studies that disagree with the conventionally deterministic working of the brain.

DEMONSTRATION OF FREEWILL

The psychiatrist, Jeffrey Schwartz (42-45.), argues that the exercise of the conscious will can overcome or reduce the problems of obsessive-compulsive disorder (OCD). This disorder leads to repetitive behaviour, for example, repeated unnecessary hand-washing. The patient is aware that their behaviour is unnecessary, but has a compulsive urge to persist with it. This behaviour is related to changes in brain function in the orbital frontal cortex (OFC), anterior cingulate gyrus and basal ganglia, all areas related to emotional processing (Schwartz 1997 a&b), (Graybiel et al, 1994), (Saint-Cyr et al, 1995), (Zald & Kim, 1996a&b). The patients are able to give clear subjective accounts of their experience that can be related to cerebral changes as revealed by scanning. Thus the sight of a dirty glove can cause increased activity in the orbital frontal and anterior cingulate gyrus. There is also increased activity in the caudate nucleus, a part of the basal ganglia that modulates the orbital and the anterior cingulate (Schwartz, 1997a, 1998a).

Schwarz states that studies of patients who learn how to alter their behaviour by the apparent exercise of their conscious will, showed significant changes in the activity of the relevant brain circuits. Anticipation and planning by the patient can be used to overcome the compulsions experienced in OCD. Patients are able to learn to change their behaviour while the OCD compulsions still occur. Successful patients are active and purposeful not passive during the process of their therapy. The actual feel of the OCD compulsions does not usual change during the early stages of treatment. The underlying brain mechanisms have probably not changed, but the response of the patient has begun to change. The patient is learning to control the response to the compulsive experience.

To make a change requires mental effort by the patient at every stage. New patterns of brain activity have to be created for the patients to be aware that they are getting a faulty message, when they get an urge to carry out some compulsive behaviour. At the same time the patients have to refocus on some more useful behaviour. If this is done regularly, it is suggested that the gating in the basal ganglia will be altered. It is suggested that the response contingencies of the TANs alter as a result of the patient frequently resisting the compulsive urges. This presumably reflects projections from parts of the cortex to the TAN cells within the basal ganglia.

What are we to make of this study, in the light of what recent neuroscience is telling us about the emotional brain? In the first place, the disorder is related to problems in the emotional areas of the brain in the form of the orbitofrontal and the anterior cingulate. These are at least in part the areas the push the patient towards hand washing or some other repetitive behaviour. However, the orbitofrontal at any rate is capable of both of changing it assessment, and evaluating the choice between conflicting rewards. As a purely speculative suggestion, I would suggest here that rational-based inputs, most likely from the dorsolateral could change the emotional weighting of particular actions so that the subjective feel-good factor of another hand washing might be balanced out by the feel good factor of overcoming the compulsion. Projections from the orbital frontal and other regions could in turn shift the balance of inputs to the TAN cells, at the rational/emotional juncture between striosomes and matrisomes.

INCREMENTAL V. ENTITY THEORISTS

Studies by the psychologist, Carol Dweck, suggest that subjects who believe they can influence their academic performance (referred to as incremental theorists) perform better than students who are convinced that their performance is preordained (entity theorists) (Dweck & Molden, 2005, Molden & Dweck). Entity theorists tend to withdraw effort and avoid tasks once they have failed. Incremental theorists attempt an improved approach to a problem task. In a study (Blackwell et al, 2007), in which entity and incremental theorists started a high school maths course with the same standards, the incremental students soon pulled ahead, with the gap continuing to widen over the duration of the course. This distinction was related to the incrementalists willingness to renew efforts after a setback. A further study (Robins and Pals, 2002) showed that during their college years, entity theorists had a steady decline in their feeling of self-worth, relative to incremental theorists. Other studies (Baer, Grant & Dweck, 2005) linked some cases of depression to self-critical rumination on supposedly fixed traits by entity theorists, and suggested that incremental theorists had greater resilience to obstacles, were more conscientious in their work, and more willing to attempt challenging tasks.

This suggests a role for conscious will or effort to act in a causal way on brains that initially had the same quality of rational problem solving so far, to leave them with different qualities by the end of a period of study. The essential distinction in the academic performance is that when the incrementalists suffered a setback, they did not accept this as the final judgement on their performance. This looks to point to a subjective assessment of two scenarios, the easy but disappointing scenario of giving up, and the demanding but more satisfying strategy of trying again. The second is absent in the entity theorists because they 'know' that they can't achieve more than a modest performance.

PROBLEMS IN CONVENTIONAL FREEWILL THEORY

The psychologist, Roy Baumeister (46.), examines the reason for the scientific and psychological consensus against the existence of freewill. He suggests a metaphysical element in this, with some scientists feeling that rejection of freewill is part of being a scientist. The fact that Libet and similar experiments have shown that actual movements of the body are not driven by free will is acknowledged, but Baumeister points to researchers such as Gollwitzer (1. 1999), who distinguishes between the decision to act and the action or movement itself. It is suggested that free will may have a role in the deliberative stage. For instance, free will could govern the decision to go for a walk, but the actions of getting up, going out the door and putting one foot in front of the other would be unconsciously driven.

Self-control, such as the ability to resist short-term benefits in favour of long-term goals and also rational choice based on deliberative thinking are here seen as two of the most important factors associated with freewill. Baumeister argues that reasoning entails at least a limited degree of freewill in that people can alter their behaviour on the basis of reasoning. Similarly self-control equates to the ability to alter behaviour in line with some goal. Decisions such as these can certainly be related to emotional evaluation in the orbital frontal and other regions.

Baumeister cautions that the ability of modern technology to study periods of milliseconds may have blinded some researchers to the importance of processes that take extended periods of time. He wonders why people agonise over decisions if they actually have no influence on them, and also suffer negative stress effects in situations where they lack control over their lives. The implication is that the use of time and energy on such a process should have been selected out by evolution if it had no relevance.

The author argues that while researchers such as Wegner have shown that people are sometimes not aware of the causes of their actions, that is very different from saying that they never determine their actions. The consensus against freewill has set the bar as high as possible in denying that freewill ever has any influence or exists at all. They have to show that none of the apparent occurrences of freewill are real, rather just producing scattered examples of freewill being an illusion, some involving rather contrived conditions.

EFFICACY AND ENERGY USE IN FREEWILL

Baumeister argues for the efficacy of freewill. In particular studies show that the processes of both self control and rational choice deplete glucose in the bloodstream, leading to a deterioration in subsequent performance. It appears unlikely that evolution would have selected for such a high energy process if it was not efficacious. Consciousness is closely associated to freewill and these studies therefore carry a strong implication that consciousness itself is also a physical thing or process involving energy and being efficacious.

In Baumeister's own experimental studies, he found that the performance of self-control tasks deteriorated if there had been previous self-control tasks. The implication of this is that some resource is used up during the exercise of self-control. The exercise of choice seems to have the same effect. Subsequent to the exercise of either self-control or choice, attempts to exercise further self-control saw performance deteriorate, in a way that did not occur when participants were just thinking or answering questions. This suggests that self-control and rational choice both draw on some form of energy. Gailliot et al (2007) (47.) found that self-control caused reductions of glucose in the bloodstream, and that low levels of glucose were correlated with poor self-control.

This finding has important implications for the freewill argument. If free choice was only some form of illusion, it is not clear why it would be adaptive for evolution to select for something that consumed a lot of energy, but had no influence on behaviour. There is a rather convoluted suggestion that we have the illusion of freewill because that makes us think that others have freewill and should therefore be punished if they do not make choices that are favourable to the group.

There are two problems with this approach. The Baumeister study showed that the same depletion of energy that occurred with the exercise of free will in the sense of self control also occurred with the exercise of choice not requiring any particular restraint on impulses. The physical process of choosing, often referring to individual or private matters looks to go far beyond simple approval of the actions of others. Further to this, if freewill is really just a charade, it is surprising that it should require such a noticeable amount of energy. In fact, the assessment of the positive or negative effects of the actions of other members of the group looks to be more easily accessible to an algorithm based process.

There is perhaps a deeper implication, not discussed in this articles that consciousness which is closely related to the experience of free choice is itself a physical thing or process requiring energy. This should not be a surprise given the nature of the physical laws, but at the moment it looks to be contrary to the scientific consensus. The high energy cost of freewill suggested here also serves to explain why conscious as distinct from unconscious processing is used only sparingly, and that is one reason why we rely on unconscious responses for much of our activities.

FREE WON'T

An area of the basal ganglia known as the subthalamic nucleus (STN) is important from the point of view of the freewill debate. Benjamin Libet (48-51.), whose experiments indicated that some minor 'voluntary' movements were initiated before subjects were consciously aware of wishing to move, postulated that there could be a 'free won't' mechanism that blocked actions that began unconsciously, but were later determined to be inappropriate by the conscious mind. Recent studies show that the subthalamic nucleus does have an inhibitory role in stopping behaviours whose execution has already begun.

SOCIETY WITH FREEWILL

The scientific consensus against freewill has created some anxiety that as this 'knowledge' gradually leaks from the laboratory into the popular mind there will be a deterioration in public behaviour. Ingenious arguments have been advanced against this, but studies suggest that we should fear such a deterioration. Vohs & Schooler (2008) (52.) found that participants who had read a study advocating the non-existence of freewill were more likely than controls to take advantage of an opportunity to cheat in a subsequent maths test. Other studies by Baumeister et al showed that participants encouraged not to believe in freewill were more aggressive and less helpful towards others.

SECTION 6

A THEORY OF CONSCIOUSNESS

At this stage, we might think we have covered enough ground to try to put together a theory of consciousness that has explanatory power, and is not obviously at variance with what we know about physics, neuroscience or evolution.

We have tried to define consciousness, as our subjective experience, or as the fact of it 'being like something' to experience things. Consciousness also involves our subjective awareness of the ability to subjectively envisage future scenarios, and to use these for our choice of actions. I have further suggested that there is only one problem with consciousness, the problem of how qualia or subjective experience arises, and that we have to address this and essentially only this in discussing consciousness.

We have examined theories of consciousness that operate within the context of classical physics, and always come up against essentially the same explanatory gap. Classical physics gives a full explanation of the behaviour of macroscopic matter, without any need for consciousness, and also without any ability to generate consciousness. This creates a problem as to how the brain can give rise to consciousness, given that neuroscience describes the brain in terms of the macroscopic matter made up of carbon, hydrogen, oxygen and other atoms, the behaviour of which can be described without either requiring or generating consciousness.

The failure to find a theory with satisfactory explanatory power within classical physics pushes us towards identifying consciousness as a fundamental or given property of the universe. What does this really mean? Explanation in science works by breaking things down into their components and the forces or processes that make them function. But this downward arrow of explanation does reach a floor. Mass, charge, spin and the particular strengths of the forces of nature are given properties of the universe that are not reducible to anything else and come without any explanation. Because consciousness has a similar lack of explanation, it is likewise suggested to be a fundamental property.

This is only a start. In itself it tells us nothing about how such a fundamental manifests in the brain. Rather than having a solution, we are only at the beginning of a very difficult journey towards something with explanatory value. Not only do we have to discover some system that is truly fundamental, but, given the lack of apparent consciousness in the rest of the universe, we need a process that is unique in operating only in brains, and not in other physical systems.

Quantum consciousness is really a misnomer for the sort of system that we are looking for. The philosopher, David Chalmers, was correct in pointing out that there was no more reason for consciousness to arise from quanta than there was for it to arise from classical structures. Both permeate the universe outside of the brain without producing consciousness. The quanta and their behaviour are only of interest if they can allow the brain access to a fundamental property not apparent in other matter.

This brings us also to the question of what really is fundamental. There are two sides to this question, the quanta or spacetime. The quanta are the fundamental particles/waves of energy, which also equates to the mass of physical objects. Some quanta such as the proton and the neutron are composed of other quanta, so are not truly fundamental or elementary. The quarks that make up the protons and neutrons of the nucleus of the atom and the force carrying particles such as photons appear to be the most fundamental quanta. But the quanta cannot be understood in isolation. They must be seen as having some form of relationship to spacetime, and that's a more difficult area than might appear at first sight.

Neither quantum theory, nor relativity which is our theory of spacetime, have ever been falsified, but they are, nevertheless, incompatible with one another. Many physicists are coming round to the notion that spacetime is not an abstraction but a real thing, and also something that is not continuous, but discrete, and perhaps best conceived in the form of a web or network. They are divided as to whether the quanta create spacetime, or spacetime generates the quanta, or the third possibility that the two are two expressions of something more fundamental. However, whatever form it is conceived to take, the concept of a real and discrete structure also allows the possibility of some form of pattern or information capable of decision making, and this is the level of the universe where we need to look for an explanation of consciousness.

There are two routes leading to the conclusion that consciousness has to derive from such a fundamental level of the universe. In addition to the view that classical physics simply can't cut it in respect of consciousness, there is the Penrose approach via the function of consciousness. As described earlier, he proposed that the Gödel theorem meant that human understanding or conscious could perform tasks that no algorithm-based system such as a computer could perform. This is led to an arcane dispute with logicians and philosophers which few lay people can follow.

However, I think it unnecessary to penetrate into such an arcane area. At a much more mundane level, the process of choosing between alternative forms of behaviour or courses of action by means of subjective scenarios of the future looks to also invoke a process that cannot be decided by algorithms. This suggestion is now supported by recent studies showing that in the orbitofrontal region the brain some activity correlates to subjective appreciation rather than the strength of signal, whereas in other parts of the brain not involved with preferences, activity correlates to the strength of this same signal. So while Penrose provides the original inspiration for the idea of an aspect of the universe that could not be derived from a system of calculations, it seems possible to simplify or streamline the original inspiration in a manner that is compatible with recent brain research, and not open to the same sort of attacks from logicians and philosophers.

In a similar way, it may be possible to simplify Penrose's proposal of a special type of quantum wave function collapse as the gateway to conscious understanding, seen here as an aspect of fundamental spacetime geometry. Penrose dismissed the randomness of the conventional wave function collapse as irrelevant to the mathematical understanding in which he was initially interested, and instead proposed a special form of objective wave function collapse, which was neither random nor deterministic, but accessed the fundamental spacetime geometry. His proposal as to wave function collapse is currently the subject of experimental testing although this is a procedure that is likely to take up to a decade.

Again the question is whether it is necessary to go to such lengths. Might there be a way around the apparent randomness that led Penrose do dismiss conventional wave function collapse. Might not the more conventional wave function collapse, or decoherence equally well provide an access to the fundamental and conscious level of the universe. There are queries as to how random the randomness is. In one form of the famous two-slit experiment, single photons arrive at a screen over some extended period of time. The initial photons register on the screen in apparently random position, but as later photons arrive the familiar light and dark bands form. Somehow later photons or perhaps the earlier photons, 'know' where to put themselves.

There is a suggestion that this puzzle links to one of the other puzzles of quantum theory, namely entanglement, by which the quantum properties of particles can be altered instantaneously over any distance. In this suggestion, the photons in the two slit experiment are entangled with other distant quanta. Whatever it is that decides the position of these particles in this scheme has no apparent explanation in terms of algorithms or systems of rules for calculating, and this is something that it holds in common with choice by emotional valuation.

But how could such a mechanism related to the fundamentals of distant space arise within our brains. Penrose's collaborator, Stuart Hameroff, proposed a scheme by which quantum coherence arose within individual neurons and then spread throughout neuronal assemblies. Most conscious commentators believe that this theory can be straightforwardly refuted because of the rapid time to collapse or decoherence for quantum states in the conditions of the brain.

However, this simplistic approach has in effect been partly refuted by the discovery of functional quantum coherence in biological systems during the last few years, initially in simple organisms subsisting at low temperatures, but most recently at room temperature and in multicellular organisms. Moreover, it is now apparent that the structures of aromatic molecules within the amino acids of individual neurons are similar to those within photosynthetic organisms now known to use quantum coherence. The structures that support quantum states in photosynthetic systems rely on the pi electron clouds discussed in earlier sections. In microtubules the amino acid tryptophan supports the same structure of pi electron clouds which thus look potentially capable of sustaining quantum coherence and entanglement through significant sections of a neuron. The mechanisms by which quantum coherence could subsist in neurons looks here to be within our grasp or understanding.

But as with the original Penrose proposal, Hameroff's scheme may be more ambitious and therefore more open to criticism than it needs to be. Where quantum states have been shown to be functional they subsist for only femtoseconds or picoseconds, whereas the Hameroff scheme requires quantum coherence to be sustained for an ambitious 25 ms, moreover it has to be sustained over possibly billions of neurons spread across the brain. This lays it open to attack from many angles.

It looks much more feasible to work from the basis of quantum coherence that exists in other biological systems and to look for similar short-lived single-cell processes in the brain. The known systems of functional quantum states that subsist within individual cells elsewhere in biology look to have the potential to exist within neurons. For this reason, it is thus much more feasible in the absence of countervailing evidence to work on the basis of consciousness arising within individual neurons. This effectively inverts the Hameroff scheme. Rather than neurons feeding into the global gamma synchrony, the synchrony, which is certainly correlated with consciousness, may be a trigger to conscious activity in neurons.

Recent studies give credibility to the idea of consciousness in single-neurons. Experimentation has shown that increased activation in single-neurons is correlated to particular conception perceptions. Some neurons are selective in only responding to particular images and activity in these is correlated with the conscious experience of particular images. Of course it isn't as simple of that. With 100 billion neurons in the brain and perhaps a good percentage of these selecting for particular images there has to be some way of coordinating their activity.

It is initially puzzling that the same type of experiments that show a correlation between consciousness and individual neurons also show a correlation between possibly billions of neurons in the global gamma synchrony and consciousness. So which produces consciousness, the individual neurons or the gamma synchrony? Recent research suggests that neuronal activity correlates with the global gamma when a number of neighbouring neurons become active together. This agrees with the processing 'hot spots' in the brain also correlated to conscious processing, which have been detected by researchers. As for the nature of the brain, the picture we are left with here is that the brain is a gate to the fundamental level of the universe. I use the word 'gate' here not as an image, but in the literal sense of a mechanism for allowing things to pass or alternatively excluding them.

All of this may seem very speculative, but it has to be remembered that this is proposed in the face of a lack of explanatory value in theories that are based on classical theory, the emerging science of quantum functionality in organisms and new indications of the physical role of subjectivity in the brain.

REFERENCES

1.) Penrose, Roger (1989)—The Emperor's New Mind—Oxford University Press
2.) Penrose, Roger (1994)—Shadows of the Mind—Oxford University Press
3.) Chalmers, David (1996)—The Conscious Mind—Oxford University Press
4.) Bohm, David (1980)—Wholeness and the Implicate Order—Routledge
5.) Schlosshauer (Ed.) (2011)—Elegance and Enigma: The Quantum Interviews—Springer
6.) Pusey, M., Barrett, J. & Rudolph, T.—The quantum state cannot be interpreted statistically—(arXiv: 1111.3328v1 [quant-ph] 14 Nov 2011
7.) Smolin, Lee (2000)—Three Roads to Quantum Gravity—Weidenfeld & Nicolson
8.) Smolin, Lee (2007)—The Trouble with Physics—Allen Lane
9.) Tegmark, M. (2000)—Importance of quantum coherence in brain processes—Physical Review E 61, pp. 4194-4206
10.) Engel, G. et al (2007)—Evidence for wavelike transfer through quantum coherence in photosynthetic systems—Nature, 446, p.782

11.) Lee, H., Cheng, Y. & Fleming, R. (2007)—Coherence dynamics in photosynthesis: protein protection of excitonic coherence—Science, 316, (2007), p. 369

12.) Collini, Elizabetta & Scholes, Gregory—Coherent intrachain energy in conjugated polymers at room temperature—Science, 323, 369

13.) Sarovar, M. et al—Quantum entanglement in photosynthetic light harvesting complexes

14.) Ishizaki, A. & Fleming, G. (2009)—Theoretical examination of quantum coherence in a photosynthetic system at physiological temperature—PNAS, 7 August 2009

15.) Cia, J. et al (2009)—Dynamic entanglement in oscillating molecules—arXiv:0809.4906v1, 29 September 2009

16.) Whaley, K. Birgitta et al (2010)—Quantum entanglement in photosynthetic light harvesting complexes—arXiv:1012.4059v1 [quant-ph]

17.) Fassiolo, Francesca & Olaya-Castro, Alexandra (2010)—Distribution of entanglement in light harvesting complexes and their quantum efficiency—arXiv:1003.3610v1 [quant-ph]

18.) Collini, Elizabetta et al (2010)—Coherently wired light-harvesting in photosynthetic marine algae at ambient temperatures—Nature, 463, pp. 644-7, doi: 10.1038/nature08811

19.) Calhoun, T.R. et al (2009)—Journal of Physical Chemistry B, 113, 16291

20.) Kauffman, Stuart (2011)—Answering Descartes: Beyond Turing—In:—The Once and Future Turing: Computing the World—Cambridge University Press

21.) Feferman, Solomon—Penrose's Gödelian argument—Dept. of Mathematics, Stanford

22.) Penrose, Roger & Hameroff, Stuart—Quantum Space-Time Geometry and Orch OR Theory—In:—Consciousness and the Universe: Quantum Physics, Evolution, Brain & Mind

23.) Bernroider, G. & Roy, S. (2005)—Quantum entanglement of K+ ions, multiple channel states and the role of noise in the brain—International Society of Optical Engineering (SPIE) vol. 5841

24.) Goodale, M. & Milner, D.—Sight Unseen—Oxford University Press

25.) Weiskrantz, Lawrence—Consciousness Lost and Found—Oxford University Press P. 26.) Melloni, Lucia et al (2007)—Synchronisation of neural activity across cortical areas correlates with conscious perception—Journal of Neuroscience, 27 (11), pp. 2858-65

26.) Melloni, Lucia et al (2007)—Synchronisation of neural activity across cortical areas correlates with conscious perception—Journal of Neuroscience, 27 (11), pp. 2858-65

27.) Singer, Wolf (2010)—Neocortical Rhythms—In:—Coordination in the Brain—Eds. Chrisopher von der Malsburg, William Phillips & Wolf Singer—MIT Press

28.) Singer, Wolf (1999)—Neuronal synchrony: A versatile code for the definition of relations—Neuron, 24, (1), pp. 49-65

29.) Tsunoda, K. et al (2001)—Complex objects represented in inferotemporal cortex by the recombination of feature columns—Nature Neuroscience, 4 (8): pp. 832-8 P.

30.) Koppel, N. et al (2000)—Gamma and beta rhythms have different synchronisation properties—PNAS, 97, (4): pp. 1867-72

31.) Whittington et al (2001)—Synaptic and non-synaptic mechanisms underlying stimulus induced gamma oscillations—Journal of Neuroscience, 21 (5): pp. 1727-38

32.) Malach, Rafael (2007)—The measurement problem in consciousness research—Behavioural and Brain Sciences P.

33.) Malach, Rafael—Weizmann Institute (2008)—Neuronal explosions: Neuronal dynamics underlying perceptual awareness in the cortex

34.) Malach, Rafael (2006)—Perception without a perceiver—Journal of Consciousness Studies, 13, no. 9. Pp. 57-66

35.) Malach, Rafael et al (2007)—Coupling between neuronal firing rate, gamma and fMRI is realted to interneuronal activity

36.) Quiroga, Q., Kreiman, G., Koch, C. & Fried, I. (2008)—Sparse but not 'grandmother cell' coding in the medial temporal lobe—Trends in Cognitive Science (2008), vol. 12, no. 3

37.) Kreiman, G., Fried, I. & Koch, C. (2002)—Single neuron correlates of subjective vision in the human medial temporal lobe—PNAS, June 11 2002, vol. 9, no. 12, pp. 8378-83

38.) Rolls, Edmund et al (2003)—Representation of pleasant and painful touch in the human orbitofrontal and cingulate cortices—Cerebral Cortex, 13, (3), pp. 308-17 doi.10.1093/cercor/13.3.308

39.) Rolls, Edmund (2005)—Emotion Explained—Oxford University Press

40.) Rolls, Edmund (2008)—Memory, Attention & Decision Making—Oxford University Press

41.) Kosiol, Leonard & Budding, Deborah (2010)—Subcortical Structures and Cognition—Springer.

42.) Schwartz, J. (1997b)—Cognitive behavioural self-treatment—In:—Obsessive Compulsive Disorder—Eds. Hollander, E. & Stern, D.

43.) Schwartz, J. (1998a)—Neuroanatomical aspects of cognitive behaviour therapy—British Journal of Psychiatry, 173 (suppl. 35), pp. 38-44

44.) Schwartz, J. & Begley, Sharon (2002)—The Mind and the Brain: Neuroplasticity and the Power of Mental Force—Harper & Collins

45.) Schwartz et al (2003)—The volitional influence of the mind on the brain—In:—Consciousness, Emotonial Sel-Regulation and the Brain—Ed. Bauregard, M.—John Benjamins

46.) Baer, J., Kaufman, J. & Baumeister, R. (2008)—Are We Free? Psychology and Free Will—Oxford University Press

47.) Gaillot et al (2007)—Self control relies on glucose as a limited energy source—Journal of Personality and Social Psychology, 92, pp. 325-336

48.) Libet, B.—Mind Time

49.) Libet, B. (2003)—Can consciousness affect brain activity?—Journal of Consciousness Studies, 10: 12, pp. 24-28

50.) Libet, B. et al—subjective referral of the timing for a conscious experience—Brain, 102, p.193 P.

51.) Libet, B. et al—The volitional brain—Journal of Consciousness Studies, 6, nos. 8/9, p.1

52.) Vohs, K. & Schooler, J. (2008)—The value of believing in freewill—Psychology Science, 19, pp. 49-54

53.) New Horizons in the Neuroscience of Consciousness (2010)—Eds.—Elaine Perry, Daniel Collerton, Fiona LeBeau & Heather Ashton—John Benjamins

54.) Dynamic Coordination in the Brain (2009)—Eds.—Christoph von der Malsburg, William Phillips & Wolf Singer—MIT Press

55.) The Orbitofrontal Cortex (2006)—David Zald & Scott Rauch—Oxford University Press

56.) Prokorny, J.—Coherent activity in the cytoskeleton— Bioelectrochemistry and Bioenergetics (1999)

57.) Prokorny, J.—Excitation of vibrations in microtubules in living cells—Biolectrochemistry, 63, (2004), pp. 321-26

58.) Guerreschi, G., Cai, J., Popescu, S. & Briegel, H.—Persistent dynamic entanglement from classical motion: How bio-molecular machines can generate non-trivial quantum states—arXiv: 1111.2126v1 [quant-ph] 9 Nov 2011

GLOSSARY

Algorithm: A procedure or rules for performing a calculation.

Amino acids: Molecules formed from carbon, hydrogen, oxygen and nitrogen. These are grouped in an amine group, a carboxylic acid group and a side chain that is different for each amino acid. Genes code for 20 amino acid and chains of amino acids form the proteins that are basic to life.

AMPA and NMDA receptors: Excitatory receptors for glutamate on post-synaptic dendritic spines of neurons.

Anthropic principle: Concept that we would not be able to observe the universe if the laws of physics did not permit the existence of intelligent life forms, and that this therefore explains why the laws of physics are to an improbable degree set just right for the existence of such life.

Artificial intelligence: Computer basis for robotics

Association cortex: Occipital, parietal and temporal cortex involved in sensory input and perception. Mainly towards the rear of the brain.

Axon: Long extension from the main body of a neuron or brain cell. Electrical signals pass down these to synapses, which communicate with other neurons.

Axon spike: Fluctuation in electrical potential in the cell membrane of the axon transmitting a signal to the synapse.

Axon terminal: The end of the axon from which neurotransmitters are sent across the intermediate synapse to the next neuron. This is the basis for information transfer in the brain.

Ca2+ channel: Ion channel in the cell membrane that selects for calcium ions.

Cartesians: Those following or deemed to be influenced by the 17th century philosopher Descartes.

Classical computers: Includes all normal existing computers, and means those computers not utilising entangled quantum states.

Classical physics: Physics before quantum theory was developed in the early 20th century and still the main way of describing the behaviour of macroscopic objects or anything much above the scale of the atom.

Coherence: (See quantum coherence)

Conformation of protein or folding of protein: Changes in the shape of proteins, which is crucial to biological processes.

Conservation of energy: First law of thermodynamics stating that energy in the universe cannot be created or destroyed.

Copenhagen theory: Early and for a long time orthodox interpretation of quantum theory, given this name because it derives from Neils

Bohr based in Copenhagen. Seeks to overcome the problems of the theory by proposing that the quanta are only mathematical abstractions until observed or measured.

Cortex: Upper and outer layers of the brain, particularly associated with processing incoming sensory signals, cognitive and some emotional processing and consciousness.

Covalent bond: Chemical bonding in which atoms share a pair of electrons. A basic bond in chemistry and the formation of molecules.

Delocalisation: Delocalised electrons are spread over several adjacent atoms rather than a single atom or the covalent bond between two atoms. A notable example is the delocalisation of the six pi electrons in aromatic molecules such as benzene and tryptophan.

Dephasing: Process by which the coherence or in-phase oscillation of the quanta breaks down and creates the classical state.

Dipole and dipole attraction: Refers to a molecule that carries opposite electric charges. Biomolecules and water are both important here.

Donor and acceptor states: Atoms providing or obtaining electrons

Dorsal and ventral stream: The dorsal stream stretches from the primary visual cortex in the occipital lobe to the parietal lobe and is involved with spatial location of objects and related actions. The extent to which it is separate from the ventral stream running from the primary visual area to the temporal lobe and responsible for visual perception is still controversial. Studies have suggested that only the processing of the ventral stream enters consciousness.

Electron cloud: Electrons around an atom are conceived as quantum waves not having a definite position, and the electrons as a whole are therefore viewed as a cloud. This is consider a more accurate view than the easier to understand 'Bohr atom' with the electrons like planets orbiting the solar system.

Electron or quantum tunnelling: As quantum waves, electrons can extend through barriers that would be impenetrable to them as particles. Such quantum tunnelling in biological matter.

Entropy: Measure of disorder. A gas concentrated in one corner of a container has low entropy. A gas dispersed evenly in a container has high entropy.

Exciton: An exciton is formed when an electron is excited out of its valence band leaving behind a positively charges whole. The exciton can transport energy through matter without transporting net electrical charge.

Femtosecond: 10^{-15} seconds or one quadrillionth of a second.

fMRI: Functional magnetic resonance imaging used to measure changes in blood flow related to neural activity. The predominant form of brain imaging since the 1990s.

GABA: Inhibitory neurotransmitter widespread in the brain.

Gamma oscillation or synchrony: Process by which electrical signalling in substantial areas of the brain oscillates in phase. Known for its correlation with conscious processing.

Gap junction (or electrical synapse): Allows direct passage of current and smaller molecules between the interiors of separate cells including neurons.

Gephyrin: Protein in the postsynaptic network of inhibitory synapses.

Glutamate: The main excitatory neurotransmitter in the brain.

Gödel theorem: Theorem proposing that with any system of axioms there would be a proposition that was obviously true but could not be proved by the axioms.

Golgi apparatus: Organelle found in cells and responsible for modifying and organising macromolecules within the cell.

GTP hydrolysis: Hydrolysis or the reduction of GTP to GDP is involved in the development of microtubules. Each of the dimers of the tubulin protein carry two GTP molecules. When these hydrolysis into GDP the microtubule can depolymerise. While the GTP remains it serves as a cap to the end of a microtubule to prevent depolymerisation.

Inferior temporal region: Part of the temporal lobe associated with object recognition.

Interference pattern: When waves intersect they form a pattern of peaks and troughs, or with light waves, light and dark bands. The appearance of this in the two slit experiment indicates the wave form of the quanta.

Ion channel: Ion channels are formed by proteins in the membranes of cells, allowing the flow of ions across the membrane. This process drives the fluctuation in electrical potential across the membrane, which is the basis of neural signalling. This refers to voltage-gated ion channels. Other ion channels respond to the binding of particular ligand molecules.

K+ channel: Also potassium channel. Ion channel selecting for potassium ions.

Lateral geniculate nucleus: Region of the thalamus relaying signals from the retina to the primary visual cortex.

Light cone: In special relativity there are both past and future light cones. The past light cone is formed by light rays converging on the observer and is the limit of the observer's knowledge of the past. The future light cone marks out the region that can be influenced by the observer.

Limbic system: Areas of the brain associated with emotional processing. Definition of these seems to have varied over time. At present hippocampus, entorhinal cortex, amygdala, orbitofrontal cortex, piriform cortex and nucleus accumbens are included.

Localisation: When a state goes from being quantum to classical the quanta become localised in a particular position. This is the condition of classical physics.

Masking: Experimental process by which an initial image is blocked from consciousness by another image.

Medial temporal region: Area of the temporal lobe around the hippocampus involved in the formation of long-term memories.

Membrane lipids: The membrane of cells including neurons are formed out of lipid layers. P. Membrane proteins: Ion channels and receptors inserted through the membrane of cells including neurons.

Midbrain: Viewed as part of the brainstem, and includes the substantia nigra, which is a dopamine producing area that interacts with the basal ganglia.

Mitochondria: Mitochondria are organelles within cells including neurons that provide the cells with energy. They convert oxygen and other nutrients into ATP. ATP releases energy via the removal of one of its phosphate oxygen groups.

Neuronal assembly: Large number of neurons acting together in neural processing.

Neuronal columns: Neurons arranged vertically through the layers of the cortex which function together.

Neurons spiking or firing: Instance when an electrical signal passes down the axon of a neuron in order to communicate with other neurons.

Neurotransmitter: Amino acid conveying messages between synapsis and dendritic receptors. Part of the basic information processing of the brain.

NMDA receptor: (see AMPA and NMDA receptors)

Non-algorithmic: Process not based on rules for making a calculation.

Non-locality: Ability of quantum entangled particles to effect one anothers quantum properties despite being out-of-range of a signal travelling at the speed of light. The effect is considered instantaneous over any distance.

Occipital cortex: Area of primary visual cortex situated at the rear of the brain.

Orchestrated objective reduction: Theory of consciousness based on the collaboration of Roger Penrose and Stuart Hameroff.

Orch OR (see orchestrated objective reduction) P. Parietal cortex: Area of the cortex handling sensory input related to spatial location and navigation.

Picosecond: 10^{-12} seconds or a trillionth of a second.

Planck length: 10-35 metres, considered to be the scale at which the continuity of spacetime breaks up. In Penrose's objective reduction the separation of superpositions of quanta become unstable and collapse above this level.

Pons structure: A part of the brain stem involved in communication between the two hemispheres of the brain.

Post-synaptic density protein (PSD): This contains hundreds of proteins including scaffold proteins and acts to ensure that membrane receptors are close to presynaptic neurotransmitters.

Primary visual cortex (V1): Located in occipital lobe at the rear of the brain, and responsible for processing signals sent from the retina via the thalamus.

Quantum coherence: The wave form of the quanta. All the possible states of the quanta oscillate in phase. Decoherence describes the process of this in-phase oscillation breaking down and giving way to a classical state.

Quantum vacuum: The vacuum as understood by quantum theory, which involves space being filled by virtual photons jumping in and out of existence, because uncertainty principle does not allow the non-existence of particles to be specified.

Receptor: Protein inserted through the neuron membrane to which neurotransmitters can bind.

Rubin vase-face: Well known ambiguous image that can be read either of a face in profile or the outline of a vase.

Schrödinger wave: Describes the development of the quantum wave. This ceases when the wave function collapses to a classical state.

Second law of thermodynamics: Entropy of a closed system including the universe can never diminish, but only either remain the same or increase.

Sensory cortex: Those areas of the cortex handling sensory input such as vision, hearing and touch.

Superior colliculus: A part of the midbrain involved in directing the eyes.

Synapses: Gap between neurons over which neurotransmitters are sent as a part of the brain's basic information processing.

Synaptic plasticity: Ability of the strength of synaptic connections to wax and wain in response to their usage.

Synaptic vesicle: Synaptic vesicles store neurotransmitters within the axon terminal for transmission across the synaptic cleft.

Thalamus: Brain region handling most incoming signals before they pass on to specific regions of the cortex.

Thermal equilibrium: Even distribution of temperature, which also indicates a high degree of entropy. Living systems are far from thermal equilibrium.

Turing machine: Refers to Alan Turing who developed the mathematical principles for computing.

Van der Waal forces: The force between dipoles on different molecules. This can be between two permanent dipoles, between a permanent dipole and an induced dipole (Debye force) or between induced dipoles (London forces).

Ventral stream: (See dorsal and ventral stream)

Ventral striatum: A part of the basal ganglia strongly connected to the limbic areas.

Virtual photon: Uncertainty principle means that the vacuum is filled by virtual photons that jump in and out of existence. Energy can convert the virtual photons into real photons as demonstrated in a recent experiment.

ABOUT THE AUTHOR

Simon Raggett developed an interest in science in his teens, but this lay dormant for many years, while he first studied history at university, and subsequently worked as an investment analyst. In the 1990s his interests began to move back towards science, and in particular the reviving study of consciousness.

In doing this, he soon became disillusioned with mainstream forms of consciousness studies, based on classical physics. By contrast, he drew inspiration from Roger Penrose's proposal that consciousness related to a fundamental aspect of the universe. However, in recent years he came to feel that the Penrose theory was itself in need of some updating, principally in response to advances in quantum biology and neuroscience. This resulted in 'Consciousness, Biology and Fundamental Physics'.